THE INTELLIGIBILITY
OF NATURE

science.culture

A series edited by Steven Shapin

Other science.culture series titles available:

THE INTELLIGIBILITY
OF NATURE

How Science Makes Sense of the World

Peter Dear

THE UNIVERSITY OF CHICAGO PRESS
CHICAGO AND LONDON

The University of Chicago Press, Chicago 60637
The University of Chicago Press, Ltd., London
© 2006 by The University of Chicago
All rights reserved. Published 2006
Paperback edition 2007
Printed in the United States of America

16 15 14 13 12 11 10 09 08 07   2 3 4 5 6

ISBN-13: 978-0-226-13948-7 (cloth)
ISBN-13: 978-0-226-13949-4 (paper)
ISBN-10: 0-226-13948-4 (cloth)
ISBN-10: 0-226-13949-2 (paper)

Library of Congress Cataloging-in-Publication Data
Dear, Peter Robert.
The intelligibility of nature : how science makes sense of the world / Peter Dear.
p. cm. — (Science.culture)
Includes bibliographical references and index.
ISBN 0-226-13948-4 (alk. paper)
1. Science—Methodology—History. 2. Science—Philosophy—History.
3. Reasoning—History. 4. Philosophy of nature—History. I. Title. II. Series.
Q175.32.R45D43 2006
501—dc22
2005026032

TO PAULINE,
MORE THAN EVER

Contents

Illustrations

Acknowledgments

Many people, both my professional colleagues and my students, have helped in bringing this book to fruition, whether by lending their specific expertise or by bringing to bear their critical acumen. Those who have read drafts of chapters and shared their thoughts and criticisms with me include Carin Berkowitz, Adrian Johns, Anna Märker, Will Provine, Lisbet Rausing, Simon Schaffer, Suman Seth, Janet Vertesi, and Heidi Voskuhl; in addition, Heidi's own work on contemporary speech-recognition software was of crucial importance in the development of some of the book's themes. Many others have heard versions of some of the book's central ideas at colloquia and conferences over the last several years, and not a few have made valuable observations and suggestions that have helped to clarify aspects of its arguments; thanks in particular to Peter Achinstein for an important reference. I am grateful to all these people for their interest and engagement.

Some of the interpretations presented in the following pages were first tried out on undergraduates in my lecture courses on the history of science. Their forbearance and willingness to entertain unfamiliar points of view—as long as they are clearly expressed and explained—underscore the intellectual benefits of the classroom.

The readers for the University of Chicago Press furnished me with two exemplary and enormously useful reports on the penultimate draft. Their diligence and commitment to the task reflect well

on the professional legacy of the late Susan Abrams, who first en-
tertained and encouraged the writing of this book; her oversight at
the press has been ably continued by Christie Henry. Finally, and
most significant of all, the critical insight of the series editor, Steven
Shapin, as well as his unfailing enthusiasm for the project, made the
whole process of its writing an intellectual pleasure. His characteris-
tically productive challenges to crucial themes or interpretations in
draft chapters of the book have also improved it considerably. With
all these debts (and with apologies to anyone inadvertently over-
looked), I can most reliably claim ownership only of remaining errors
of fact and interpretation.

As usual, I dedicate this little book to Pauline, who keeps my spirits up and draws diagrams.

My gratitude for support in the writing of the book goes to the
National Endowment for the Humanities (FA-36050-00), and the
John Simon Guggenheim Memorial Foundation.

Peter Dear
Ithaca, NY
19 Jan. 2005

Science as Natural Philosophy, Science as Instrumentality

I. Two Faces of Science

What do you do when you want to know about some aspect of the natural world? Most people, certainly most people in the industrialized world, would find out what the scientists have to say. If you want to know about the stars, you ask an astronomer or an astrophysicist; if you want to know about biological inheritance, you ask a geneticist; if you want to know about the history of the earth, you ask a geologist or a geophysicist.

"Science," taken as a general category, is a very prestigious label that we apply to those bodies of knowledge reckoned to be most solidly grounded in evidence, critical experimentation and observation, and rigorous reasoning. Science is practiced, as a matter of circular definition, by scientists. Despite the diversity of specialized scientific disciplines, you may be sure that, even if the first professional scientist of whom you ask your question is of the wrong specialty, that person will guide you to another scientist (or something written by one) who actually is an expert in the relevant field—scientists are recognizable as a group by their tendency, in such circumstances, to stick together. And from these people you will receive an account of how things work, or how things are, in the natural world around us—an account of what kind of universe it is that we are a part of.

In giving their accounts, the scientists will be telling you about what used to be, but is no longer, called "natural philosophy." That term largely fell into disuse during the nineteenth century, but in the first half of that century and earlier it was the standard way of referring to an intellectual endeavor aimed at understanding nature. By the end of the nineteenth century, natural philosophy had become absorbed into "science" in the sense in which we know it today. But the term "natural philosophy" is perhaps one worth reviving, precisely because it emphasizes that aspect of science which is concerned with explaining and understanding the world—what is often called the "scientific worldview." When journalists and popularizers treat such figures as Albert Einstein, or Stephen Hawking, or Stephen Jay Gould as people with profound insight into the true nature of the universe, or when they tell us what scientists have to say about the fundamental structure of matter, or about distant galaxies, or life on Mars, this is the face of science that they present: science as natural philosophy, which strives to give an account of nature—to make sense of it. But, of course, this is not the only face of science.

Science is also about power over matter, and, indirectly, power over people. Scientists are not only fonts of wisdom about the world, our "priests of nature," typically inhabiting universities; they are also people who work for business corporations and for military concerns (interests frequently shared, indeed, by the universities themselves, with their business and government contracts). Scientists and their science, in other words, do practical things that others want. The popular image of a scientist is of someone in a white coat who invents something—a vaccine, a satellite, or a bomb. Indeed, the main reason for the great prestige of science seems to be that the word is frequently associated with technological achievement. Alongside science as natural philosophy, therefore, we have science as an operational, or instrumental, set of techniques used to do things: in short, science as a form of engineering, whether that engineering be mechanical, genetic, computational, or any other sort of practical intervention in the world.

Indeed, when scientists are invited to testify before congressional committees in the United States, or otherwise to provide advice to those in political power, their status as authorities, as experts, resides above all in their presumed ability to pronounce on matters of pressing practical importance. Whether these matters concern the assessment of environmental risks associated with industrial pollution or the health risks associated with dietary or drug regulations, people with doctorates in relevant scientific fields are regarded as the ones most fit to provide guidance, because scientists, on this standard view, *know how nature works.*

However, a number of recent scholars argue that an easy and direct association between scientific truth-claims and technical achievements is much less obvious than is usually supposed. Cases of the direct "application" of so-called "basic" or "pure" science, when examined closely, show that the practical and theoretical effort that scientists have to exert in order to get things to work properly is much greater than the usual distinction between "pure" and "applied" science would suggest. In fact, as the history of science shows time and again, it is sometimes unclear that the world even contains the natural objects referred to by the theory supposedly being "applied."

II. Instrumentality and the History of Science

If we thought that successful application demonstrates the truth of relevant theories, we would have to believe that all of space is filled with a material, mechanically structured substance called "aether," which occupies even those regions that we normally think of as being completely empty. This aether would be composed of particles that move in particular ways so as to produce the forces found in the phenomena of electricity and magnetism. It was, after all, on the basis of such a picture that James Clerk Maxwell, in the 1850s and '60s, first developed the theory that predicted the existence and the means of producing radio waves. But despite the fact that radio waves were originally predicted on the basis of Maxwell's theory (by the German physicist Heinrich Hertz in the 1880s), few people would nowadays want to say that the technical ability to produce and detect them means that there is truly a mechanical aether filling the universe.

When we look back over the history of science, we do not see the clear, progressive development of a single picture of what the world is like, of what kinds of things it contains and the ways by which they interact. Instead, we see a picture that changes constantly in many of its most prominent features. To take a particularly striking example: before the acceptance of Einstein's special theory of relativity in the early twentieth century, many physicists believed, like Maxwell, in a material aether as the medium that carried the forces of light, electromagnetism, and even gravity. After the acceptance of Einstein's theory, however, the aether had simply vanished; it was no longer needed. Far from being gradual and cumulative, the shift amounted to a radical alteration in views of what the world is like.

There are many other examples of this sort, such as ideas about the true nature of heat. In seventeenth-century Europe, during the so-called Scientific Revolution, heat came to be regarded as an effect of particles of matter in rapid agitation. In the eighteenth century, that idea was generally replaced with the view that heat is a kind of fluid, called caloric, that pervades bodies like water in a sponge. That theory fitted in well with new eighteenth-century ideas concerning specific and latent heats, and with chemical theories of matter. It proved so useful that it was not abandoned until the middle of the nineteenth century, after which the idea of heat as a rapid motion of particles returned to favor—the kinetic theory of heat. The history of science is full of flips back-and-forth on fundamental questions about the underlying nature of physical phenomena.

This fact does not imply that nothing really changes in scientific (natural-philosophical) understanding. Obviously it does; the kinetic theory of heat that was developed in the nineteenth century was useful in accounting for a lot that the much vaguer seventeenth-century concepts about heat as motion could not. But ideas about what kind of thing in the world corresponded to the multifarious phenomena of heat (as with electromagnetism after Maxwell) clearly did not proceed in any clear, cumulative line of development. If there was "prog-

ress," that progress certainly did not take the form of an ever-closer approximation to a true picture of what heat really is.

Why, then, should any particular view of the nature of heat (or,
for that matter, radio waves) be preferred over any other? A standard
answer to such questions is that scientific theories are believed to be
true because they work; philosophers sometimes speak of the practical success of science as something to be explained by the truth of
its theories: in effect, practical efficacy is used as evidence for the
truth of the natural philosophy which, it's assumed, underlies and
explains it. But the "efficacy" of Maxwell's original theory of the
electromagnetic aether is no longer taken as evidence for the real existence of the aether; instead, different theories of electromagnetism, which posit different constituents of the universe, are now attributed the efficacy that Maxwell's theory was once supposed to
possess. At the same time, many theories are still employed for the
purposes of practical engineering that are no longer believed to be
literally true in their natural-philosophical content—a simple example being the use of earth-centered astronomy for purposes of
navigation. Either way, the practical efficacy of scientific theories,
what can be called their "instrumentality," is a component of science
distinguishable from its natural philosophy.

Just as this instrumentality is routinely assumed to provide support for natural-philosophical assertions, so arguments of a natural-
philosophical kind are often used to explain the instrumental success
of particular techniques. The wave properties of electrons, a part of
the nãtural philosophy presented by quantum mechanics, are used
to explain how and why electron microscopes work; DNA profiling
is an effective technique because people believe that the natural philosophy of modern genetics and molecular biology explains and
justifies it. If the natural philosophy were not believed, the technique
would not be seen as effective.

That certain techniques really are effective is clearly the case, and
is one of the defining features of modern industrial and postindus-
trial societies. But to imagine that the efficacy attributable to modern
science flows directly from the truth of its representations of the

world, that is, from its natural-philosophical content, is unrealistic. It would do a grave injustice to the work and intellectual content of technical and engineering practices. Such accomplishments, frequently and routinely attributed to something called "science," in fact result from complex endeavors involving a huge array of mutually dependent theoretical and empirical techniques and competences. There is usually only a tenuous path back to a natural-philosophy component located amidst the tangle. The "D" in "R&D" (research and development) is not, as any practitioner will tell you, a trivial exercise in "application" of theory; there will always be jobs for test pilots.

Nonetheless, the widespread and longstanding assumption that science's instrumentality is nothing more than a matter of "applying" the knowledge provided by its natural philosophy has had an enormous cultural impact. The authority of science in the modern world rests to a considerable extent upon the idea that science is powerful; that science can do things. Television sets, or nuclear explosions, can act as icons of science because it is taken for granted that they somehow legitimately represent what science really is—the instrumentality of science, that is, often stands for the whole of science. At the same time, when science is appealed to as the authority for an account of how something really is in nature—that is, when science is seen in the guise of natural philosophy—its accepted instrumental efficacy seems to justify that image of truthfulness.

Science in its broad "umbrella" sense is an amalgam of natural philosophy and instrumentality, even though each is not always clearly present with the other in every field: astrophysicists or cosmologists do not yet offer the prospect of manipulating black holes, for example. But the overall totalizing effect of the amalgam is hugely powerful. Why are science's instrumental techniques effective? The usual answer is: by virtue of science's (true) natural philosophy. How is science's natural philosophy shown to be true, or at least likely? The answer: by virtue of science's (effective) instrumental capabilities. Such is the belief, amounting to an ideology, by which science is understood in modern culture. It is circular, but invisibly so.

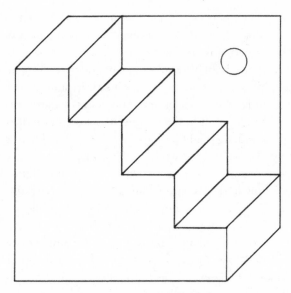

0.1. An optical puzzle: is the wall marked with a circle behind a staircase, or in front of a ter-raced ceiling?

The way in which these two faces of science are related is not, then, a straightforward matter of two different definitions sitting side by side, always sharply distinguished from each other. In fact, each of those two categories is what is known as an "ideal type": a coherent system of representing some kind of thing in the world. One might, for example, want to describe a society as fundamentally communist, or free-market capitalist, or fascist, even though any real example will always have idiosyncrasies that make it depart to a greater or lesser extent from the ideal type that best describes it. The complication in the case of science is that it best corresponds not to one, but to *two* ideal types simultaneously. A useful analogy is the well-known kind of visual illusion that can be seen in either of two ways: does the staircase in the picture (fig. 0.1) go upwards from right to left, so that the wall with a circle on it is at the back, or does this show a terraced ceiling, the circle being on the nearest wall? Most people can switch between seeing the picture in one of those ways and in the other—but not both at once. And yet the picture itself is always unchanged; it is not the case that it switches back and forth. Similarly, "science" can be represented in modern culture in its guise as natural philosophy or in its guise as instrumentality, but not both at the same time. When a scientific statement is regarded as a piece of natural philosophy, it has the status of a description of the natural world. Something might perhaps be done with it, but as it stands, it is simply about how the world is. Conversely, when a scientific statement is regarded as an expression of instrumentality, it is an account of how to do something, an account that may also be said to have implications about how the world is. Scientific statements are inherently ambiguous as to which sort they claim to be, natural-philosophical or instrumental. But although the ambiguity is seen in terms of just one way or the other, *both* perceptions must always be at least conceivable—even if astrophysicists might have difficulties in actually manipulating a galaxy.

This ambiguity suggests that the widespread modern view of science is in important ways fallacious; that it attributes to science a character that it does not in fact possess. If that is so, how could things have come to be that way? What sustains the illusion? An answer will require an excursus into the history of science, and in particular into the history of one half of the twin components of modern science, natural philosophy.

III. Natural Philosophy and Intelligibility

Natural philosophy is a much older enterprise than modern science. Within broadly European intellectual traditions, from antiquity to the seventeenth century, the term itself denoted an enterprise that was later associated with that of "science," as that category took full shape in the nineteenth century. But in medieval and early-modern Europe, natural philosophy was understood in literate culture as an enterprise that was entirely separate from practical knowledge, or know-how. The distinction was rooted in the works of the Greek philosopher Aristotle (fourth century BC), which formed the backbone of university education throughout the entire period: Aristotle's Greek terms were *epistēmē* and *technē*, translated into Latin in the Middle Ages as *scientia* (science) and *ars* (art). The first item in the pair designated logically and empirically demonstrable knowledge of truth, while the latter referred to the skilled practice of manipulating material things (*technē* is a root of the word "technology," just as *ars* is of the word "artificial"). The reason for Aristotle's making such a sharp distinction between the two is generally reckoned to reflect ancient Greek social arrangements: only free men, such as the citizens of the city-state, had the leisure to devote their time to philosophizing, while practical abilities were the province of servants and slaves. Not surprisingly, therefore, a free man like Aristotle regarded *epistēmē* as being a much worthier and more noble object than the skills of the manual laborer, and of a quite different character. The clerical, contemplative world of the medieval and early-modern university evidently agreed; it adopted Aristotle's categorization wholeheartedly. And the particular *scientia* concerned with understanding and explaining the natural world was "natural philosophy," often called "physics" (from the Greek *physis*, a word that stands for "nature").

A good example of what this sense of natural philosophy meant is the distinction that was made throughout this period between the natural philosophy of the heavens, on the one hand, and astronomy, on the other. Natural philosophy as it concerned the heavens was all about such questions as what the heavens are made of, what moves the sun, moon, and planets, whether the universe is finite or infinite; questions to do with understanding the *nature* of the heavens. Astronomy was something entirely different: for one thing, it was not part of natural philosophy. It counted instead as a part of mathematics, an applied branch that had to do with the positions and motions of celestial bodies—tracking lights in the sky and applying geometry to their behaviors. All that astronomy aspired to do was to provide formalized descriptions and predictions of heavenly motions, for practical uses—casting horoscopes, constructing calendars, or navigating by the stars, for example. Astronomy declined to ask what the moving bodies actually were, or why they moved, because trying to understand the heavens was not its concern; it just charted them. Understanding was the job of natural philosophy, which in turn had no interest in crunching numbers for utilitarian ends.

This widespread view of natural philosophy began to change during the seventeenth and eighteenth centuries. During that time there emerged a new view of natural philosophy whereby it would have, as one of its primary goals, practical uses. This new conception was especially associated in the seventeenth century with the ideas of Francis Bacon (1561–1626) and in the early eighteenth century with the work and propaganda of Isaac Newton's English followers and publicizers. Bacon had actively opposed the established Aristotelian conception of natural philosophy, and asserted that a true natural philosophy should be concerned with active intervention in nature for the benefit of mankind. Bacon presented this project as one of Christian charity, to alleviate suffering. Because he was a lawyer and statesman, Bacon viewed these matters as a proper concern of the state itself, the most obvious, and powerful, agent for the amelioration of the human condition. Despite his political failure to get his ideas enacted as English government policy, Bacon's writings on the subject struck a chord that resonated throughout Europe, and they represent a widespread ambition in this period among the educated. By the eighteenth century, following the work of Newton (1642– 1727), such views had become almost commonplace. Newton's followers in England, as well as others elsewhere in Europe, promoted a view of nature that underpinned and legitimated a much greater concern with practical matters on the part of the upper echelons of society. These were especially people who sought to make money from improvements in agriculture, mining, and, increasingly, manufactures. A natural-philosophical universe like Newton's, which consisted of lifeless matter bouncing around according to mechanical laws, was clearly exploitable for human ends: it lent itself to instrumentality. Tinkering with the world-engine could now be justified as natural philosophy, not artisanal labor, although some still complained that the distinction between the two was being ignored, and that Newton's work was in fact *not* real natural philosophy at all.

Doing things and understanding things thus became increasingly folded into one another. Eventually, the label "natural philosophy" itself faded away, absorbed by "science": Michael Faraday (1791–1867), in the mid-nineteenth century, was resisting a trend when he insisted on being known, still, as a "natural philosopher," and yet he involved himself in a good many practical scientific enterprises, from chemistry to telegraphy.

When we restrict the label "natural philosophy" to its original sense, that of intellectual understanding of the natural world, we can see clearly that the cultural activity called "science" as it developed during the nineteenth and twentieth centuries has not been the same as the old natural philosophy. The changes that the latter label had undergone during the seventeenth and eighteenth centuries resulted in the establishment of a new enterprise that took the old "natural philosophy" and articulated it in the quite alien terms of instrumentality—science was born a hybrid of two formerly distinct endeavors.

The natural-philosophical component of science is the one that has, perhaps, the most profound role in shaping our views of ourselves. The universe in which we live, the bodies that we experience as part of ourselves, and the sense we have of our immediate environments are all shaped by our acceptance of the images of reality that we owe to science in its guise as natural philosophy. But what is it that gives scientific knowledge-claims their powerful authority over our imaginations? One answer is, of course, their association with instrumentalities of various sorts, which is why it is now fashionable in some academic quarters to refer to contemporary science and technology as "technoscience," representing the two as a single enterprise. That association appears endlessly in popular science fiction. At a more fundamental level, however, there is the powerful social authority of science, which serves to render most people unable to refuse a knowledge-claim presented as a "scientific fact," even though they are incapable of judging its truth for themselves. The academic credentials of scientists and the institutions at which they work are important parts of the general credibility of science.

There is more to the natural philosophy inside science than just facts about what is in the world, and how it behaves, and where it came from. Besides facts, there are explanations and stories that pull the facts together into accounts that make sense. Making sense, in turn, means more than just conforming to the rules of logical inference. Examples that are considered in later chapters of this book include such epoch-making work as that of Isaac Newton on gravity. When Newton first published his great work, the *Principia*, in 1687, with its assertions and demonstrations concerning the mutual gravitational attraction of matter and the inverse-square law governing the relation between the force of that attraction and the distance between the gravitating bodies, many critics objected that, for all the impressive mathematics in the book, what Newton had done was not natural philosophy. For those critics, a true natural-philosophical, or physical, explanation of gravity would require an account of what gravity "really was," rather than just a description of how it behaves. They also knew what kind of account would do. For them, a physical explanation of gravitational attraction would be a strictly mechanical one, by which they meant the effect of direct contact-action of matter pushing on matter. And Newton had been quite explicit about not providing one—because, try as he might, he was unable to come up with a mechanical explanation that would fit his mathematical description of gravity, a description that accorded so well with empirical measurements.

The reason for this general demand for a "mechanical" explanation of gravitational attraction is that a good number of prominent natural philosophers had, during the course of the seventeenth century, become convinced that only explanations for natural phenomena that were couched in the terms of lumps of dead, inert matter pushing against other lumps of inert matter, and thereby pushing them, were truly *intelligible*. Newton appeared to many people to be suggesting that gravity operated by strict action at a distance: a material body simply exerted a force on another body that was separated from it by empty space, with nothing passing between them. But Newton's critics thought that this just made no sense. And since Newton had not explained gravitational attraction in a way that made sense, he had failed as a natural philosopher—although not as a mathematician.

During the course of the eighteenth century, however, many people became accustomed to the idea of action at a distance. As they became used to working with it, and applying Newton's principles to the solution of specific, mathematically structured problems, philosophers ceased to worry about the foundational objections that had earlier been leveled against it. The result was that action at a distance eventually came to be seen, almost by default, as an explanatory approach that was satisfying to the understanding. In the nineteenth century, the situation again shifted. James Clerk Maxwell (1831–79) and other British physicists in the second half of the century argued that a satisfactory explanation of gravity (as well as of other phenomena) could be couched only in terms of contact-action using, as we saw earlier, a mechanical aether. Ironically, both Maxwell and Michael Faraday cited Newton as a supporter of their view that anything else would be implausible at best (see chapter 6).

Another striking example of intelligibility, again from the nineteenth century, concerns one of the most profound shifts of recent centuries in sensibilities about how nature should be understood. Charles Darwin (1809–82) found the idea of natural selection a satisfactory way of explaining how the organic world comes to display so much apparent designfulness and purpose—how animals and plants seem to be so well fitted to the kinds of lives they lead. Rather than infer, as most of his predecessors had done, that the functional *designfulness* of living things implied a benevolent Creator God who had deliberately made them that way, Darwin suggested that, had they not been as they are, they would not have been able to survive: the living beings that we see around us are the ones that happened to have what it took to survive—they had, as it were, won life's lottery. But for some of his contemporaries, this "natural selection" was inadequate to its explanatory task; for them, it just wasn't the right *sort* of explanation. The doyen of mid-Victorian philosophy of science, John Herschel (1792–1871), famously described natural selection as "the law of higgledy-pigglety." He said this not to pour scorn on it—he was certainly able to follow the argument—but because random, statistical processes seemed by definition unable to account meaningfully for order in the universe; apparent design surely required real intelligence (see chapter 4). His response parallels Albert Einstein's (1879–1955) celebrated condemnation of the probabilistic, rather than strictly causal, explanations of quantum mechanics: "God doesn't play dice with the universe." Evidently, then, there are no timeless, ahistorical criteria for determining what will count as satisfactory to the understanding. Assertions of intelligibility can be understood only in the particular cultural settings that produce them.

"Intelligibility" is ultimately an irreducible category—by definition, you cannot analyze fundamental, bedrock principles down to anything more basic than themselves. An account makes sense just because it does, not because of some prior condition or criterion: the intelligible is the self-evident; the unintelligible is simply the unspeakable. In the historical development of science, the awkward and unresolved tension between instrumentality and natural philosophy has yielded views of the universe that are dependent on particular human conceptions of what makes sense. Presumed intelligibility is an essential ingredient of natural philosophy, and in that sense natural philosophy is, and always has been, about feeling at home in the world. Perhaps the difficulties that some people find in feeling at home in the modern world are at least in part due to the way that science's instrumentality has increasingly displaced part of that natural-philosophical intelligibility.

CHAPTER ONE

The Mechanical Universe from
Galileo to Newton

I. The World as a Machine

The seventeenth century is often described as the period in which the enterprise of modern science was born. The so-called Scientific Revolution, on this view, saw the birth in western Europe of many of the characteristic features of science as we know it today: intellectual, social, and institutional. But those latter two categories, the social and the institutional, probably owe their typical modern forms (especially the conduct of scientific research in universities) to precedents set in the nineteenth rather than the seventeenth century. Nonetheless, in the intellectual arena the importance of what happened in the seventeenth century is undeniable, because this period saw the emergence of one of the most dominant metaphors in subsequent science: the metaphor of the world as a machine.

There is nothing inevitable about seeing the world as a kind of machine. On the contrary, in cultures the world over, including most of earlier European history, the commonest way of seeing the natural world has been in terms of living organisms. In the learned culture of premodern Europe—the culture of universities and of books—the most influential version of this view derived from the writings of the ancient Greek philosopher Aristotle. Aristotle based his approach to nature on the ways in which we tend to make sense of living processes around us. Plants and animals, as well as people, are born, grow, become old, and die. Aristotle found it natural to model his understanding of all processes in the world on these kinds of experiences. For him, therefore, behaviors in the natural world were made intelligible by understanding them as processes, directed towards a goal. An acorn sprouts, sets roots, and grows. Why? Because it is on its way to becoming an oak tree. A dog runs towards a piece of meat. Why? Because it wants to eat it. Purpose, goals, make sense of all kinds of processes in the world, just as they do in explaining the behavior of human beings, and for Aristotle these goals did not even have to rely on conscious intentions. The acorn growing towards becoming an oak tree is not conscious of an intention to do so; it just does it. But we still explain what happens in terms of how the process will usually (barring accidents) end up. This form of explanation is known in English as *teleological,* from the Greek word *telos,* meaning goal.

Teleology characterized all of Aristotle's universe. It even explained such things as why stones fall to the ground: they do so because they are seeking the center of the universe (where Aristotle located the earth itself). This was the way in which the world of late medieval Europe, with Aristotelian philosophy entrenched in its universities, organized its natural philosophy and made sense of the world. Things and occurrences were intelligible when they could be understood in terms of processes that aimed at some purpose.

This was the natural philosophy that many educated Europeans in the seventeenth century rebelled against. The major names in philosophy and the sciences in that period, such as Descartes, Galileo, and Newton, began to criticize the idea that teleological explanations were appropriate for understanding nature, and advocated in their place explanations that privileged mechanical causation. The model would no longer be a growing organism, but a clock. By knowing the arrangement of the clock's component parts and the ways in which they pushed against each other, one could understand the characteristic movements displayed by the clock's hands. To explain the movements of the hands by reference to their purpose of displaying the right time seemed to be ridiculous. In fact, an Aristotelian philosopher would not have disagreed on that point. He would simply have regarded it as inappropriate to use a clock as the model for explaining *natural* phenomena. But for people like Descartes, even the growth of a plant was something to be understood in terms of inert matter in motion, analogous to clockwork.

The seventeenth-century conflict between Aristotelian philosophers and the new proponents of what was called the "mechanical philosophy" is an excellent example of radically differing views of scientific intelligibility in conflict. Each group (despite wide differences among individuals in the same camp) sought natural-philosophical understanding, but they could scarcely agree on what true understanding meant. For one group, explanations of natural phenomena in terms of mechanical interactions failed to make sense of the very processes supposedly being explained, whereas for the other group, explanations of natural phenomena in terms of teleology themselves failed to make sense. Were inanimate objects to be ascribed souls that could have desires and intentions? Aristotelians did not believe that nonliving things required souls or minds in order for their behaviors to be explained teleologically, but mechanists claimed not to understand how goal-directed behavior made any sense without them.

One of the clearest ways to see the difficulties faced by both sides is to look at how they criticized one another. The rules of engagement in this dispute were not set out clearly in advance, and the victory of one side or the other could not be determined to everyone's satisfaction by the application of formal logic. Furthermore, what was at issue had nothing to do with disagreements over what phenomena there were in the world to be explained; empirical investigation would not settle matters. In that sense, this was a fundamentally philosophical debate, and specifically a natural-philosophical one. It deeply concerned the nature of the universe, rather than resting on the affirmation or denial of controversial physical phenomena.

So a typical procedure in attacking a philosophical opponent in the seventeenth century was to ridicule him. Here is an example from one of the writings of René Descartes (1596–1650), one of the most influential of all the mechanical philosophers of the seventeenth

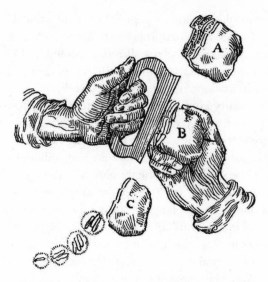

1.1. From Descartes's Principles of Philosophy *(1644): a human mechanical interaction with matter—hitting a stone to produce sparks—illustrates Descartes's mechanical ideas about the nature of sparks, which are created from the tiny spheres initially embedded in the rock (top).*

century. In this work, written in French in the early 1630s and called *The World*, Descartes provided a detailed sketch of the entire physical universe, with every last feature of it to be explained in the mechanical terms of lifeless bits of matter pushing against one another, much as a human worker creates effects by pushing matter against matter (fig. 1.1). In the course of the account, Descartes pauses to criticize the standard Aristotelian explanation of what motion really is:

> And trying to make it more intelligible, they [i.e. Aristotelian philosophers] have still not been able to explain it more clearly than in these terms: *motus est actus entis in potentia, prout in potentia est*. The terms are so obscure to me that I am compelled to leave them in Latin because I cannot interpret them. (And in fact the words "motion is the act of a being

which is in potency, in so far as it is in potency" are no clearer for being in the vernacular.)[1]

Descartes wants to reject the Aristotelian explanation of motion not on the grounds that his is preferable, or that empirical evidence disproves theirs, but simply on the grounds that their explanation *makes no sense*. And he does this by ridiculing them. He behaves as though he is unable to show the insufficiency of the Aristotelian account of motion by analyzing it and showing it to be faulty, or self-contradictory; instead, he appeals directly to his reader's intuitions of intelligibility.

In some circumstances, this tactic might well appear as a "know-nothing" mockery of things that the ridiculer does not himself properly understand. But Descartes, no fool and also no ignoramus, surely understood what he was attacking. In short, Descartes made fun of this piece of Aristotelian philosophy because he thought that its faults were so deep-rooted that they could not be corrected; there was no basis on which to make sense of it. So he said so.

Descartes's pretended inability to understand the meaning of this explanation of motion was, of course, not one that was shared by Aristotelian philosophers themselves. They believed that they understood it and that it was a good explanation of the phenomenon. Descartes too would, we may reasonably suppose, have been capable of restating and elaborating on the sentence that he quotes in much the same language as its supporters would have done; he would not have been dumbfounded by it. But he clearly felt that however the explanation was interpreted, it would always come down to the acceptance or rejection of something that for him simply *did not make sense*. Any argument about the matter was always going to end in deadlock.

Descartes played on a common expectation, of course: his confession of ignorance was evidently tailored for a reader who would have sympathy with his position. Someone who speaks from a weak or inferior position can scarcely boast of an inability to understand the opponent's arguments, and certainly not in so comprehensive a manner. Perhaps the most lucid exemplar of this kind of approach is the great polemicist Galileo Galilei (1564–1642). He used a similar ploy in a book devoted to supporting the doctrine of a moving earth, the *Dialogue Concerning the Two Chief World Systems* of 1632. Galileo's reader will quickly have learned that this dialogue's fictional character Salviati, who attacks Aristotle and defends Copernicus's doctrine of a moving earth, is to be applauded, while the character Simplicio, a caricature of an Aristotelian, should be seen as foolish. So when Simplicio purports to explain why bodies fall by reference to their *gravity*, Salviati replies by ridiculing the use of the word as an *explanation*. What is it that moves earthly things downwards? "The cause of this effect," says Simplicio, "is well known; everybody is aware that it is gravity." Salviati responds in this way: "You are wrong, Simplicio; what you ought to say is that everyone knows that it is called 'gravity.' What I am asking you for is not the name of the thing, but its essence, of which essence you know not a bit more than you know about the essence of whatever moves the stars around."[2] Similarly, when the English philosopher Thomas Hobbes, a few years later, ridiculed the usual Aristotelian explanation of fall as an endeavor towards the earth's center, he wrote: "As if Stones, and Metalls had a desire, or could discern the place they would bee at, as Man does; or loved Rest, as Man does not; or that a peece of Glasse were lesse safe in the Window, than falling into the Street."[3]

A good case can be made, however, that Aristotelian explanations like this one, involving "qualities" such as gravity that were possessed by things, were by no means as empty of content as Galileo or Hobbes tried to make out. For example, an explanation involving the quality "gravity" (i.e., "heaviness") as something that a body can possess proposes the existence of a kind of thing in the world, a real quality, that does not exist according to other ways of explaining fall. The Aristotelian position, that is, does not simply play with words, but makes claims about what sorts of things the world contains. Nonetheless, Galileo had no more fear than Descartes of being held up to ridicule by a competent philosopher, even though he misrepresented the point of his opponents' arguments. Galileo's rhetorical position was easily strong enough, because the audience that mattered was already on his side. The common assumption of Galileo and the rest seems to have been that, since (according to them) the Aristotelian arguments make no sense, an unflinching consideration of those arguments in the very words in which they are expressed will make that unintelligibility self-evident. In order, then, to convict Aristotelian explanations of failing to make sense, all that these writers can do is to pillory them; there is no other recourse.

In 1620 the English statesman and philosopher Francis Bacon (1561–1626) had written in his work *New Organon*, defending his own critique of Aristotelian philosophy, "There is no easy way of teaching or explaining what we are introducing; because anything new will still be understood by analogy with the old."[4] He had earlier made a similar point: "No judgement can rightly be made of our way (one must say frankly), nor of the discoveries made by it, by means of *anticipations* (i.e. the reasoning currently in use); for one must not require it to be approved by the judgement of the very thing which is itself being judged."[5] Bacon tried, that is, to prevent people from rejecting his innovations before he had a chance to present them properly. The difference in position between Bacon on the one hand and Galileo or Hobbes on the other is that the latter writers were on the offensive, and did not need to clear space to allow them to develop their own ideas. They wanted their opponents to be branded as *wrong* on the grounds that the arguments of those opponents made no sense, and they expected their readers to share their views. (They were also sometimes the victims of similar tactics.)

Negative arguments could only go so far, however, and Descartes in particular had his own especially systematic ways of explaining natural phenomena. The technique that he employed both in *The World* and in his later, fuller account of his world system, the *Principles of Philosophy* (published 1644) was the curious and interesting one of telling an explicitly fictional story about the origins of the universe.

This story, which Descartes called a "fable," was not intended to be believed. It was meant to persuade his readers that the account he gave of how the universe currently works was a plausible one.

> I want to wrap up part of it [i.e., the account] in the guise of a fable, in the course of which I hope the truth will not fail to manifest itself sufficiently clearly, and that this will be no less pleasing to you than if I were to set it forth wholly naked.[6]

He then goes on to a "Description of a new world, and the qualities of the matter of which it is composed,"[7] so as to show that this new world could be understood perfectly by his reader, and that, furthermore, it would be indistinguishable from the real world. Descartes explained this strategy more fully in the *Principles* and stressed even more strongly that his description was not of the actual world at all, and that this approach had close parallels with accepted ways of investigating nature:

> I shall even make some assumptions which are agreed to be false.
>
> Indeed, in order to provide a better explanation for the things found in nature, I shall take my investigation of their causes right back to a time before the period when I believe that the causes actually came into existence. For there is no doubt that the world was created right from the start with all the perfection which it now has. The sun and earth and moon and stars thus existed in the beginning, and, what is more, the earth contained not just the seeds of plants but the plants themselves; and Adam and Eve were not born as babies but were created as fully grown people. . . . Nevertheless, if we want to understand the nature of plants or of men, it is much better to consider how they can gradually grow from seeds than to consider how they were created by God at the very beginning of the world.[8]

Descartes will thus propose basic principles, or "seeds," to provide explanations that show how the things in the world *could* have come into being.

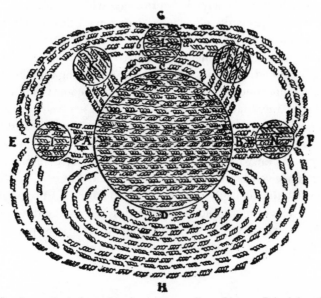

1.2. The micro-mechanical operation of a magnet according to Descartes, Principles of Philosophy: *tiny screw-shaped particles circulate, like bodies in a whirlpool, around and through the substance of the magnet, which has developed rifled pores through which the particles can pass. The handedness of the screws, whose mechanical properties are (supposedly) no different from those of normal-sized hard material bodies, explains the distinction between the two poles of a magnet.*

For although we know for sure that they never did arise in this way, we shall be able to provide a much better explanation of their nature by this method than if we merely described them as they now are. And since I believe I have in fact found such principles, I shall give a brief account of them here.[9]

He then proceeds to give an account of the universe in which everything is explained in terms of inert pieces of matter interacting with one another by collision or direct pressure (fig. 1.2). Matter itself he characterized as nothing but geometrical, or spatial, extension, because he thought that this was the only property of matter that was necessary to it, and therefore represented what matter truly, *essentially*, was. As he wrote elsewhere, "I described this matter, trying to represent it so that there is absolutely nothing, I think, which is clearer and more intelligible."[10] The result of Descartes's "fable" was, he claimed, "a world in which there is nothing that the dullest minds cannot conceive, and which nevertheless could not be created exactly the way I have imagined it."[11]

So intelligibility, for this founder of the mechanical worldview, was the primary criterion for judging an explanation of natural phenomena. As for what made something "intelligible," or capable of being conceived by the "dullest minds," that was obvious: it was simply that the intelligible idea be "clear and distinct." If it was, then not only was it intelligible, but it was also (because God would not deceive us) true as well. Descartes thought that the perception of intelligibility took work; it was so easy to go wrong.

II. Mechanical Intelligibility

Other mechanical philosophers later in the seventeenth century usually shied away from the inference that intelligibility implied truth (Descartes himself often hedged on the question). They tended instead to say that we can never know whether any explanation that we invent is really true; the best we can do is to say that it accounts for the phenomena in an intelligible way. And such people, like Descartes, thought that mechanical accounts were the clearest and most intelligible kind.

Among the more important mechanists of the second half of the seventeenth century were Robert Boyle (1627–91), in England, and the Dutchman Christiaan Huygens (1629–95). Huygens was, with the exception of Isaac Newton (1642–1727), the leading mathematical physicist of the later seventeenth century. Both Huygens and Boyle often described mechanical explanations as being peculiarly "intelligible"—that was the main reason for preferring them over alternatives. The point was not that there was any guarantee that mechanical explanations were *true;* instead, the chief virtue of them was their intelligibility itself.

Huygens's theory of gravity, presented in 1669 to the Royal Academy of Sciences in Paris, provides a good example. Huygens accounted for the tendency of heavy bodies to move towards the center of the earth by imagining countless tiny particles circling around the earth in all possible planes and in all directions. These particles would naturally tend to push outwards, centrifugally, like a stone swung around on the end of a cord. What prevented them from actually flying away was the surrounding medium, a special kind of matter that filled interplanetary space. Objects made of ordinary matter, which was heavier and more sluggish than the tiny orbiting particles, did not share the particles' rapid motion about the earth's center. Consequently, those objects did not possess the centrifugal, outward tendency of the tiny particles; they tended instead to be pushed *towards* the center by the particles' tendency to push *away* from it. That was the explanation of why ordinary bodies are heavy and tend to fall downwards, an explanation based on an idea of Descartes's.

Huygens called this explanation of gravity a "hypothesis." Descartes had always sought explanations that would be, as far as possible, certain and true, but Huygens's mechanical explanations sought not truth, but *intelligibility* instead. Huygens was convinced that a mechanical philosophy of the kind advocated by Descartes, although it could never be proved to be the necessarily true account of the universe, was in any case the only one capable of providing explanations that were truly intelligible—that is, explanations that made perfect sense. Huygens thought that his hypothesis of the cause of gravity accounted for the phenomena as least as effectively as any other known account; its superiority lay in the fact that it did so in terms of a world picture that included nothing but inert matter in motion. Huygens regarded any physical account that violated the explanatory limits of that world picture as incapable of making sense.

The paper Huygens read to the academy in 1669 began:

To find an intelligible cause of gravity, it's necessary to see how it can be done while supposing [i.e., postulating] in nature only bodies made of the same matter, in which no quality is considered, nor any inclination for each to approach the others, but only different sizes, shapes, and motions.[12]

This was an expression of the mechanistic language of intelligibility. Any particular mechanical explanation of a phenomenon might or might not be true, but any purported explanation that was not mechanical (in Huygens's understanding of that term) would be fundamentally unintelligible.

Boyle, Huygens's English contemporary, was explicit on the same point: although mechanical explanations should be preferred for their intelligibility, their intelligibility did not imply that such explanations were necessarily *true*. It was entirely possible that God might bring about phenomena by some mysterious nonmechanical means; He was, after all, omnipotent and capable of miracles as well as of immaterial, spiritual interventions. But any explanation that was nonmechanical would be philosophically unintelligible to human beings, and therefore would contribute nothing to the development of a natural philosophy. Boyle's intention was to understand the natural world, not to treat it as a great divine mystery.

The dominance of the mechanical philosophy in the later seventeenth century strongly affected the reception of Isaac Newton's work on gravity. Newton's *The Mathematical Principles of Natural Philosophy* first appeared in 1687. Originally published in Latin (and usually known as the *Principia*), which made it more accessible to a learned international readership, Newton's book nonetheless met with considerable resistance among philosophers in Continental Europe. This was not due to its mathematics. Indeed, this was among its great strengths. The English philosopher John Locke (1632–1704) wrote to Huygens not long after the *Principia* was published to ask him about the soundness of the mathematics in the work, which was beyond Locke's capacity. Huygens told Locke, who professed himself interested in the nonmathematical, philosophical content, that the mathematics *was* the *Principia;* that was the work's real content. Such a view expressed Huygens's most charitable perception of Newton's achievement: Newton's mathematical descriptions of the motions of planets, comets, and moons within the solar system, and his correlation of those motions with the behavior of heavy bodies in the vicinity of the earth's surface, represented a great mathematical achievement. The postulation of an attractive force acting between heavy objects according to an inverse-square law enabled this very economical account of motion—but was it anything more than mere description? Huygens thought not. He was a natural philosopher, while Newton himself described the *Principia* as providing the mathematical principles "of natural philosophy." But natural philosophy, Huygens insisted, was properly about providing causal explanations, whereas Newton offered no *explanation* of gravitational attraction—let alone a specifically mechanical one.

Newton was sensitive to these sorts of criticisms because he too had the expectations for mechanistic intelligibility shared by his contemporaries. It was reckoned by some in the 1690s that he was asserting that gravitational attraction was a true action at a distance between separated but mutually gravitating bodies, an interpretation to which Newton reacted with scrupulous alarm. He wrote to a correspondent: "Pray do not ascribe that notion to me, for the cause of gravity is what I do not pretend to know and therefore would take more time to consider of it." And in a subsequent letter to the same person, he stated forthrightly:

> That gravity should be innate, inherent, and essential to matter, so that one body may act upon another at a distance through a *vacuum*, without the mediation of anything else, by and through which their action and force may be conveyed from one to another, is to me so great an absurdity that I believe no man who has in philosophical matters a competent faculty of thinking can ever fall into it.[13]

The absurdity was a question of the simple unintelligibility of action at a distance: how *could* a body exert a force on another unless something moved between them or mediated between them as a means of communication? This point, to Newton, was a sufficient reason for rejecting the idea. Accordingly, at other times he postulated the existence of subtle, space-filling matter to mediate (inadequately) gravitational effects, or even suggested that they might be due to the direct action of God.

III. Problems with Mechanism

The acceptance of this mechanistic way of speaking about the intelligibility of nature obviously related to quite basic ways of experiencing the material world—a world in which people make things happen by pushing and hitting and manipulating bodies. Mechanical devices like watermills or clocks displayed those kinds of interactions very clearly and, unusually, represented cases where human beings (or their animals) were not performing the basic actions that did the work. By the late seventeenth century in literate Europe, those were the sensibilities that dominated much of natural philosophy. At the same time, natural philosophers knew that however much they implied (more often than explicitly asserted) the straightforward intelligibility of mechanical explanations of nature, such explanations were far from being philosophically unproblematic. Indeed, during the course of the century a number of basic difficulties had been found, difficulties that might well leave mechanical explanations with as many foundational problems as Aristotelian physics was held to possess.

One of the most fundamental of such problems concerned the way in which the motion (or pressure) of one piece of matter could be understood as causing motion in another piece of matter. This process was absolutely crucial to any attempt at mechanical explanations of natural phenomena, of course: such explanations used as their basic constituents the pushing of matter against matter to produce a resultant effect. Nonetheless, this simple, and apparently clear, causal process could not easily be shown to make sense philosophically. The crucial property of solid bodies that was needed here was impenetrability, and Descartes in particular was hard-pressed to find a way of giving that property to material bodies as a "clear and distinct" attribute of what it meant to be such a body in the first place.

One of Descartes's correspondents brought up the issue: the English theologian Henry More (1614–87) noticed that Descartes had tried to define material bodies solely in terms of their extension in space. Indeed, for Descartes spatial extension, the taking up of room, was essentially all that bodies really *were:* matter was space, and space, by definition, was matter—because extension is the only property that we can see "clearly and distinctly" to be necessarily and invariably associated with our idea of body. How, More wanted to know, can simple extension have in addition the property of impenetrability, of hardness?

Descartes tried to argue in response that an extended body must by its very nature be impenetrable. If, he says, one body were to penetrate another, then in the common region in which they coexist (overlap), the total amount of space that those overlapping portions had previously occupied would now be reduced by half—in effect, $1 + 1 = 1$. And since space *is* material body, that would imply the annihilation of matter in the interpenetrating bodies. "But what is annihilated does not penetrate anything else," says Descartes; "and so, in my opinion, it is established that impenetrability belongs to the essence of extension"[14]—since penetrability would lead to a logical self-contradiction.

This was all very well as an attempt to save Descartes's position, and it in fact shows clearly one of the features of "science as natural philosophy": it would not have been sufficient simply to say that matter-as-extension is *in fact* impenetrable, and leave the explanation of that fact as a mystery of no practical importance. Failing to explain things was a *philosophical* fault, given Descartes's grandiose claims; "matters of fact" would have been adequate if nothing but practical rules of use were involved, but here, instead, the point was to explain how things are and how they make sense in the world. If Descartes had been unable to do that, he would have failed at his self-appointed task.

Similar problems confronted other mechanists later in the century. Huygens, for example, attempted to lay down rules governing collisions between (imaginary) perfectly *hard* bodies, by which he meant bodies that would rebound from each other—as opposed to cases of collision between bodies like cotton balls, which lose their force of motion when they strike. Huygens thought that perfectly hard bodies (think of completely incompressible billiard balls), when they run into each other, ought to rebound at the same speed as they had previously approached one another, except that this time it would be a speed of separation instead of approach. This was a useful ideal case (although finding real bodies that would exhibit this behavior exactly would be another matter). On its basis, together with a general principle of relativity of motion, Huygens was able to elaborate an impressive set of rules governing collisions between both equal and unequal bodies moving with various initial velocities.

But the value of his rules for performing calculations was not enough to make them the foundation of a mechanical natural philosophy until Huygens had achieved a full understanding of what happened *during* a collision, and why. When two perfectly hard, incompressible bodies collide with one another and rebound, as in Huygens's ideal cases, what makes them rebound at all?

This proved to be a difficult problem. First of all, the bodies were assumed to be perfectly hard, that is, incompressible. So when one hit another, neither would deform in shape even slightly; the collision would be a matter of one contacting the other, in the case of spheres at a mathematical point, followed by an instantaneous rebound. What caused the rebound? A true "mechanical philosopher" like Huygens could not argue that the colliding balls somehow actively repelled one another with some immaterial force, because one of the central points about mechanical explanations was that they explained solely in terms of *inert* matter, that is, dead, *inactive* matter that cannot in itself do anything. Another, obvious attempt at a solution was to imagine that the perfect rebounds occurred, not between absolutely hard bodies, but between elastic ones that compressed on impact and then expanded, like springs, to throw the colliding bodies apart again. But, unfortunately, that solution too was unsatisfactory, because it simply pushed the problem back a further step, rather than solving it. *Why* would a compressed body spring back again? It wasn't easy to say and still keep the answer within the bounds of strict mechanical intelligibility.

There was always, of course, a pragmatic solution to this kind of problem. Robert Boyle himself, who liked to stress the intelligibility of mechanical explanations, took the pragmatic route in making sense of the compressibility of air (as in "Boyle's law"). He referred to the "spring of the air" as a way of talking about the tendency of compressed air to push outwards again, thus comparing it to a metal spring being compressed; his onetime assistant, Robert Hooke (1635–1702), who was involved in the work relating to "Boyle's law," also came up with "Hooke's law," concerning the direct proportionality of compression or bending to the force or weight exerted on the spring. In other words, Boyle and Hooke were sometimes prepared, despite their self-professed "mechanical" propensities, to accept certain phenomena as givens rather than to insist upon fundamental, mechanical, causal explanations of them (see fig. 1.3). At those times, they appear to have relied on the practical experience of manipulating objects as their touchstone for determining an explanation's intelligibility. Not everyone, that is, saw matters in the same way.

IV. Newton, Mechanism, and Explanation

These issues could have significant implications for natural philosophy. Perhaps the most spectacular instance concerns the natural philosophy of Newton, a figure who (among other things) is notable for his attempts to integrate mathematical reasoning into a natural-philosophical account of the world. The particular mechanical problem that evidently gnawed at Newton arose from the same problems of collision and transfer of motion that Descartes and Huygens had

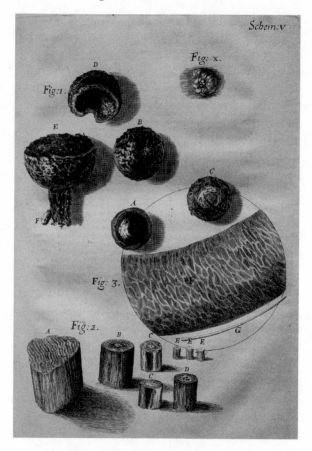

1.3. From Hooke's Micrographia *(1665), showing the tiny metallic spheres created by the sparking of a steel by hitting it with a piece of flint.*

worried about. Huygens had distinguished between collisions involving hard bodies that rebounded without losing any motion, and soft bodies that collided and thereby lost motion. Already, prior to Newton's work, philosophers had noticed that a system of the world of the kind developed by Descartes, in which the entire universe was made up of bodies continually colliding with one another, would

tend over time to run down. The total amount of motion in such a
universe could never be increased, since the interactions between its
constituent inert bodies could only at best conserve their combined
quantity of motion. At worst, in the cases of collision between soft,
nonrebounding bodies, the motions producing the collisions could
be destroyed altogether. So, overall, the universe ought inevitably to
lose motion over time, eventually becoming stagnant.

For Newton, as for others in this period, this conclusion simply
would not do. A universe that ran down was in no way a stable universe, but would be inevitably degenerating, and that was no universe for an Almighty Creator to have made. Natural philosophy, as
we have seen in the cases of Descartes and Boyle, frequently involved
considerations of God: God was the fundamental underpinning of
everything that existed, so the universe was not simply Nature; it was
also God's Creation. That was the perspective from which Newton
decided that a strictly mechanical universe failed to yield a satisfactory natural philosophy. As he wrote in the third edition of his *Opticks*
in 1717:

> Motion is much more apt to be lost than got, and is always upon the De-
> cay. For Bodies which are either absolutely hard, or so soft as to be void
> of Elasticity, will not rebound from one another. Impenetrability makes
> them only stop. If two equal Bodies meet directly *in vacuo,* they will by
> the Laws of Motion stop where they meet, and lose all their Motion, and
> remain in rest, unless they be elastick, and receive new Motion from
> their Spring. . . . If it be said, that they can lose no Motion but what they
> communicate to other Bodies, the consequence is, that *in vacuo* they can
> lose no Motion, but when they meet they must go on and penetrate one
> another's Dimensions.[15]

Newton had a solution, but, ironically, it was not a mechanical solution in Descartes's sense at all.

> Seeing therefore the variety of Motion which we find in the World is al-
> ways decreasing, there is a necessity of conserving and recruiting it by

active Principles, such as are the cause of Gravity, by which Planets and
Comets keep their Motions in their Orbs, and Bodies acquire great Mo-
tion in falling; and the cause of Fermentation, by which the Heart and
Blood of Animals are kept in perpetual Motion and Heat . . . and the
Sun continues violently hot and lucid, and warms all things by his Light.
For we meet with very little Motion in the World, besides what is owing
to these active Principles.[16]

Newton's idea of "active principles" was his solution to the problem
of the inevitable loss of motion in a strictly mechanical universe.
In effect, he took the intelligibility of Cartesian-style mechanical explanation as the starting point in his search for a better kind of natural philosophy. But having once identified precisely where that kind
of intelligibility *failed* to yield a satisfactory natural philosophy, Newton introduced active principles to provide everything that passive,
inert matter could not.

Locate two massy bodies at rest some distance apart in the far-off
voids of space, and they will begin to accelerate towards each other.
So, argued Newton, gravity must be due to an active principle. The
sun is not seen to burn out, so presumably it too keeps going by active principles. No matter that Newton could not say what these
principles were; he knew that they must exist if any sense was to be
made of the phenomena.

These Principles I consider, not as occult [i.e., hidden, imperceptible]
Qualities, supposed to result from the specifick Forms of Things, but
as general Laws of Nature, by which the Things themselves are form'd;
their Truth appearing to us by Phænomena, though their Causes be not
yet discover'd. For these are manifest Qualities, and their Causes only
are occult.[17]

Newton is concerned here to distinguish himself from the Aristotelians, who served by this time (the early eighteenth century) as the
whipping boys of people trying to appear philosophically astute.
Newton was himself criticized by important philosophers for al-

legedly setting up gravitational attraction as a similarly "occult" quality, so it was important for him to deflect such accusations:

> And the *Aristotelians* gave the name of occult Qualities, not to manifest Qualities, but to such Qualities only as they supposed to lie hid in Bodies, and to be the unknown Causes of manifest Effects. . . . Such occult Qualities put a stop to the Improvement of natural Philosophy, and therefore of late Years have been rejected.[18]

Newton's natural philosophy was a careful exploitation of widely utilized ways of presenting natural philosophies throughout the course of the seventeenth century. The mechanical philosophy had rested on the claimed superior intelligibility of mechanical explanations for natural phenomena. Its promotion involved at the same time the denigration of Aristotelian philosophy, its most important rival, on the grounds that it used *un*intelligible explanatory resources. When Newton introduced marked deviations from the norms of mechanical explanation by introducing active principles, he therefore attempted to preserve the appearance of philosophical orthodoxy by attacking the usual suspects.

Newton's targeting of Aristotelian occult explanations, although common enough by this period, is a telling one. "Occult" meant literally "hidden," and causes that are hidden were exactly the opposite of intelligible causes. The point of mechanical explanations was that they provided causal accounts of why certain phenomena occur that claimed immediate intelligibility; Boyle and Huygens both stressed that advantage to the virtual exclusion of questions of truth. Mechanical explanations were for them simply the best way to do natural philosophy. By contrast, as Newton said, Aristotelian "occult Qualities put a stop to the Improvement of natural Philosophy," and that was why they had been rejected. Newton almost implied that mechanical properties, by contrast, had *promoted* its improvement—except that, had he done so, he would have had to acknowledge explicitly that his active principles were not strictly mechanical.

Newton wove a delicate path through what he could do, on the one hand, and what he wanted to claim he was doing, on the other. He produced elaborate formal descriptions of the gravitationally mediated motions of heavy bodies in the solar system, including at the surface of the earth; he also exhibited properties of light and the differential refractions displayed by different spectral colors. In the case of the first achievement, his work received general acceptance, and in the case of the second, despite a good deal of initial resistance, his work received a serious hearing and eventual acceptance. But Newton's characterization of his work as natural philosophy required him to hedge it round with caveats and qualifications.

For a good many natural philosophers around 1700, and despite Newton's claims to the contrary, the contents of the *Principia* did not amount to natural philosophy. Christiaan Huygens's position, as already mentioned, was rooted in a firm mechanistic understanding of what natural philosophy should be. Huygens accepted that Newton had demonstrated the existence of inverse-square-law gravity as a phenomenon, but the task in natural philosophy was to explain that phenomenon—and for Huygens that meant to explain it in strict mechanical terms.

Newton knew that he was unable to do that, but he defended his work in the *Principia* as "natural philosophy" nonetheless. Just as he did in the passages from his *Opticks* quoted above, he said that, while the causes of gravitational attraction remained unknown, nonetheless he had demonstrated to a near-mathematical certainty the *existence* of universal gravity, and that itself should be good enough to count as natural philosophy. He made his argument by stressing, not the content of his claims about nature (which, after all, lacked the specification of causes that Huygens thought necessary), but the way in which his results were achieved—in other words, he effectively tried to define natural philosophy methodologically. Again, he explained it most straightforwardly in the *Opticks:*

[T]o derive two or three general Principles of Motion from Phænomena, and afterwards to tell us how the Properties and Actions of all corporeal Things follow from those manifest Principles, would be a very great step in Philosophy, though the Causes of those Principles were not yet discover'd: And therefore I scruple not to propose the Principles of Motion above-mention'd [he means the laws of motion first proposed in the *Principia*], they being of very general Extent, and leave their Causes to be found out.[19]

Newton argued that natural philosophy was a matter of establishing knowledge of natural effects (here, the "Principles of Motion") in the *right way;* it need not necessarily specify the causes of those effects as well, desirable though it would be to do so. Knowledge of effects could be useful, of course, but as such it had not usually been regarded as a part of natural philosophy properly so-called; instead, it was descriptive knowledge, or what was often called in the seventeenth century "natural history." Natural history did not involve understanding the things that were known, and it was therefore no sort of philosophy. And as something of potentially practical value, it also held a lower intellectual status.

In the eighteenth century, Newton's universe of matter in motion and forces, like gravity, acting at a distance became accepted as itself a mechanical worldview. But that acceptance implied a change in the meaning of "mechanical." For Descartes and others in the seventeenth century, "mechanical" referred to inert matter that could interact with other matter only through direct contact, whether through continuous pressure or by collision. Newton's mysterious and seemingly occult gravitational forces became incorporated into the "mechanical" as the result of familiarity and usefulness; the more that gravitational forces were used to compute results concerning the behaviors of masses moving through space (however theoretical many of those calculations might be), the less mysterious those forces seemed. The fact that no one claimed properly to understand them worried fewer and fewer people, because their instrumental role took up the slack. Newton's methodological criterion for the conduct of natural philosophy disguised this crucial issue beneath the growing view of natural philosophy as an endeavor of discovery, an enterprise whose philosophical conduct was more important than its philosophical results.

The interweaving of natural philosophy and instrumentality in Newton's work took the form, therefore, of attempts to reduce the one into the other. Instrumentality could *become* natural philosophy, Newton hoped, if natural philosophy was characterized by its methodology; but at the same time he also wanted to retain a connection to a form of natural philosophy that spoke about the true natures of things, and about how the universe really is. He allowed himself to do this publicly only in carefully cordoned-off situations where he could speculate without opening himself up to the risk of being shown to be wrong. Thus in the *Opticks* he presented in the bulk of the work his mathematical-experimental demonstrations concerning what he regarded as the manifest properties of light, which were to be accepted regardless of what one thought light itself might actually be. Then, at the conclusion of the *Opticks,* Newton appended a section containing a number of "Queries." In these openly speculative paragraphs, he did indeed put forward suggestions as to what light, as well as various other aspects of nature such as gravity, might truly be. But for Newton, the mere intelligibility of a causal account could not guarantee its truth, nor even, as it did for Huygens, restrict acceptable hypotheses to those that were based on contact-action mechanism. The conflation of natural philosophy with instrumentality was now well under way.

CHAPTER TWO

A Place for Everything:
The Classification of the World

I. Everything in Its Place

Mechanical intelligibility was never an inevitable way of understanding nature, as the Aristotelian opponents of men like Descartes show clearly. Aristotle understood what a thing was when he could say what kind of thing it was. It makes sense for a dog to eat raw meat, because, after all, it's a *dog;* that's the sort of behavior that those kinds of things typically display. Similarly, it makes sense that a stone falls to the ground when it is released, because it's a heavy body (described by Aristotle as being made primarily of the element "earth"), and that's what heavy bodies, by definition, do. Once you know what a thing *is,* you can explain many aspects of its properties and behaviors. So knowing how to classify and name things functioned for the Aristotelian as an alternative means to that of mechanical explanation for rendering them intelligible—of fitting them comfortably into one's mental world.

Making sense of the world by means of seeing it as a great machine failed to grasp *why* the things in the machine-like world were the way they were. Newton could say how a particular mass would behave in the presence of other masses, but that did not explain why one mass was a rock, another a camel, and another a petunia.

The problem of classifying the sheer diversity found in the natural world was one that fascinated natural philosophers in the century after Newton. It was not a fascination that signaled a return to the natural philosophy of Aristotle, but it did continue, in a different guise, some of the themes that had dominated Aristotelian natural philosophy before its challenge by the mechanical philosophers of the seventeenth century. Taxonomy (from the Greek word *taxis*, meaning "arrangement") was in part an enterprise that aimed at getting inside the natures of things, and understanding the interrelationships that made sense of them; it was not restricted to neutral description. During the eighteenth century, classification became increasingly employed as a route towards natural-philosophical knowledge in a wide variety of subject areas.

In the early years of the century, some natural philosophers began to approach the developing science of chemistry by constructing tables that would interrelate certain kinds of reactions in such a way as to permit systematic predictions. The real pioneer in this work was a Frenchman, Étienne François Geoffroy (1672–1731), who published an influential paper in the 1718 *Mémoires* of the Royal Academy of Sciences, in Paris. This paper included a table in which were listed, along the top row, various substances (salts in solution, common acids), and in the columns below them a number of substances such as metals. The order in which the substances appeared depended upon the particular substance at the head of the column, and represented the relative tendencies for each to dissolve in that substance at the expense of those below it in its column: if a metal some way down the column were dissolved in its particular designated salt solution or acid, and a metal listed above it in that column were then added, the new metal would itself dissolve, while the previously dissolved metal would come out of solution. So the order of the substances in any given column enabled a direct inference to be drawn concerning their behavior in relation to one another when added to the substance designated by that column; the order was not the same in every column (see fig. 2.1).

One of the most notable features of Geoffroy's paper was its refusal to speculate upon the reasons for these differential solubilities. Geoffroy simply spoke of the differing *rapports* exhibited by each metal

TABLE DES DIFFERENTS RAPPORTS
observés entre differentes substances.

Mem. de l'Acad. 1718. Pl. E. pag. 212.

2.1. Geoffroy's table of rapports, in Mémoires of the Royal Academy of Sciences, 1718.

for the substance in which it dissolved; those metals higher in the column simply had a greater *rapport* for the substance than those below them, which was another way of saying that they displaced them from solution. The table itself, therefore, was worth publishing regardless of its causal explanation—Geoffroy had made a systematic discovery, but he made no claims to a causal understanding of it.

Isaac Newton also has a role in this story, since he too had published a discussion of such displacement reactions. In the second edition of his *Opticks* (1706), he described a similar set of relationships between metals that displaced one another from solution in a variety of solutes. Unlike Geoffroy's later (and apparently independent) presentation of such material, however, Newton tried in the *Opticks* to suggest a possible explanation of this behavior in terms of differential short-range attractive forces operating between the particles of the various substances; indeed, the proposed explanation was Newton's only reason for introducing the material in the first place. His idea was that the fundamental particles of the various substances involved in the reactions exerted forces of differing intensities upon one another. For example, iron displaces copper, while copper displaces silver, from solution in *aqua fortis* (the acid formed by adding niter [potassium nitrate] to oil of vitriol [sulfuric acid], which corresponds to modern nitric acid). Newton suggested: "does this not argue that the acid Particles of the *aqua fortis* are attracted . . . more strongly by Iron than by Copper, and more strongly by Copper than by Silver," and similarly for other metals?[1]

Also unlike Geoffroy, Newton did not list these relationships in tabular form. There is something about Geoffroy's paper that places the emphasis not on the natural-philosophical understanding of chemical phenomena (and, more generally, of the particulate nature of matter), as Newton had, but on the brute phenomena themselves. Geoffroy had trained in pharmacy and medicine, and was a practicing physician, as well as professor of medicine at the Collège Royale in Paris; practical considerations of *materia medica* were frequent concerns of his work. He was criticized by others, however, for the theoretical emptiness of his 1718 paper, and responded in another paper of 1720 in which he made attempts to account for some of the phenomena in terms of the shapes of constituent particles of the substances involved, in what was by this time a conventional French Cartesian fashion.

Geoffroy's lead was not taken up in a serious way by other chemists until the 1750s. The Frenchman Pierre-Joseph Macquer (1718–84) had published in 1749 a book with the telling title *Elements of Theoretical Chemistry* (*Élémens de chymie-théorique*), in which he reprinted Geoffroy's original table. Macquer described the relationships in the table, which Geoffroy had called "rapports," as "affinities," and he discussed various potential modifications and additions to the table.

Other chemists who adopted this approach were Guillaume François Rouelle (1703–70) and his pupils (numbered among whom was the great Lavoisier). Several of them worked at developing fuller and more elaborate versions of Geoffroy's table, for a wide range of reactions (although Rouelle himself continued to specialize in the chemistry of salts). Most significant of all was their adoption of Geoffroy's own original refusal to engage in interpretation or explanation of the phenomena that they represented in their tables. They designated the topic of their tables "elective affinities," analogous to Geoffroy's impressionistic talk of *rapports*, leaving aside attempts to account for the features of the tables. The tables ordered phenomena as plain, irreducible facts—much like the facts of conventional natural history, which dealt with species of plants and animals. The structures of the tables appeared to show regularities in the relationships between individual chemical substances and between particular reactions; it was these structures themselves that were the philosophical content of the work, not some supposed explanation in terms of the physical properties of the material substances. Distance-forces acting between particles of matter, or interactions between the characteristic shapes of material corpuscles, played little part in the chemistry of elective affinities. The tables represented a way of understanding chemical phenomena—just as taxonomic charts in botanical or zoological classification were regarded as potential ways to organize and understand the world of living things, to perceive their relationships as God had established them. One of the touted virtues of the chemical tables was that they might also serve to predict the results of hitherto untried reactions, in addition to those on the basis of which the particular table was arranged. This aspect of the tables was sometimes described in explicitly natural-philosophical terms, as a means of making chemistry chime with the understanding, not simply a means of making useful predictions. The permanent secretary of the Royal Academy of Sciences, Bernard de Fontenelle (1657–1757), had this to say concerning Geoffroy's work:

> This table becomes in some sort prophetic, for if substances are mixed together, it can foretell the effect and result of the mixture, because one will see from their different relations what ought to be, so to speak, the issue of the combat: which will overcome these, which will yield to those, which will be at length victorious.[2]

Such an achievement held promise for the development of physical science in general:

> If Physics cannot reach the certainty of Mathematics, at least it cannot do better than to imitate its order. A Chemical Table is by itself a spectacle agreeable to the Mind, as would be a Table of Numbers ordered according to certain relations or certain properties.[3]

Rouelle attempted, in work of the 1740s on salts, to organize the field through a conventional form of taxonomy similar to that of the naturalists. His basic classificatory technique involved consideration of the structure of the crystals of chemically similar kinds of salt. He divided all so-called "neutral" salts (that is, salts that were neither acidic nor alkaline) into six sections on the basis of the shapes of their crystals. The use of crystal form as a classificatory criterion in the chemistry of salts became a standard technique in the second half of the century.

II. Natural History and Natural Philosophy

Botany provided the model for classification in the eighteenth century. One of the imperatives for the development of complex taxonomic schemes in botany was practical: ever since the opening up of the world to Europeans, with the now-routine voyages across the Atlantic Ocean and to India and China, naturalists in Europe were increasingly confronted with previously unknown kinds of plants that had no precedent in the writings of ancient and medieval naturalists. The explosion in the number of new species seemed positively to *require* some kind of orderly classification scheme.

But this apparent requirement cannot explain the proliferation of classificatory schemes of all sorts that the eighteenth century produced. First of all, even in the case of the new plants, the only real practical incentive to paying close attention to them at all came from their potential medical and agricultural benefits. Nonetheless, naturalists were worried about classifying *all* plants, not just those with possible practical value. The same thing happened, on a reduced scale, in the case of animals, which did not figure prominently in the armamentarium of physicians. All sorts of variety in nature seemed, to the philosophers of the eighteenth century, to need orderly classification schemes to relate them together in some appropriate way, to put similar things nearby one another and to separate things that seemed very different from each other. During the course of the century, classification schemes were devised for many different categories of things, including diseases (nosology), rocks (mineralogy), and even society (the origins of the national census lie in this period, as the Constitution of the United States bears witness).

For natural philosophers in the century after Newton, knowing how to classify or name something correctly was a way of understanding it; classification was itself a form of knowledge. What made this knowledge into true natural philosophy (as opposed to being merely knowledge of a human convention, such as an alphabetical listing) was the assumption that the way in which things were organized into groups reflected something about their actual relationships in nature. The most elaborate classification systems were developed in botany, due to the large numbers of species involved. Botanical taxonomies had already become fairly elaborate by the end of the seventeenth century, but the issues involved in their justification were by no means settled. One of the major taxonomists of the late seventeenth century, the English naturalist John Ray (1627–1705), argued in the 1690s against his rivals (most of whom believed otherwise) that man-made taxonomic schemes for organizing plants or animals were ultimately only arbitrary, and in fact were unable to reveal genuine relationships in nature itself. Ray's reasons for holding this position are revealing of the grounds for the differing convictions of his taxonomic opponents.

Ray understood that a truly natural classification of plants or animals would be one that grouped individual species together on the basis of their true affinities. If, for example, two species of plant were to be grouped together as members of the same genus (the immediate higher-level category), that should ideally mean that those species were truly similar to one another *in the order of nature*. Plant specimens that looked almost identical to one another except for some minor but characteristic difference (say, in the exact appearance of some feature of their flowers) might readily be classified as different species belonging to the same genus, but the greater such differences became, and the more numerous they were, the less obvious was the choice of appropriate characters that should be used in classifying the plants. Ray used a particularly striking example from zoology. Should, he asked, the whale be counted as a creature most closely related to warm-blooded land animals, such as horses and cows, or should it be classed as a kind of fish? It all depends on which of its characters are taken as the most important. Whales possess fins, live in the sea, have warm blood and give birth to living young (rather than eggs). Those first two features might serve to link them to fish, whereas the second two link them to the land animals (the category and term "mammal" had not yet been invented). So which of the available characters should be chosen in deciding on the correct category for whales? Indeed, what would it mean to speak of a "correct" as opposed to an "incorrect" classification at all?

Ray did not, however, mean to suggest that the choice of characters to use in classification was a purely arbitrary matter. He thought that there were indeed some characters that were appropriate to use as criteria of classification and some that were inappropriate. But he argued that it is not possible to tell which is which. God knows, but human beings cannot read God's mind. As a result, we are reduced to choosing our classificatory criteria, our selection of some characters of an organism in preference over others, on a purely pragmatic basis: we pick those characters that, on the whole, seem to do the best job of grouping similar species together and grouping dissimilar species apart from one another, in separate categories. There was no short-cut available that would allow the identification of particular characters that guaranteed correct taxonomic groupings—we do not know what such characters, or, indeed, such groupings, are; we can only guess.

Most of Ray's colleagues and rivals in natural history balked at such a pessimistic conclusion, even while recognizing the practical difficulties inherent in taxonomy. The underlying reason for their reluctance seems clear: admitting to the imprecise and pragmatic basis of taxonomy in natural history would reduce that activity to a trivial descriptive exercise—as we might say nowadays, to a form of stamp collecting. Most of Ray's fellow naturalists wanted to see themselves as natural philosophers, which meant that natural history had to be portrayed as a philosophical activity that sought the true natures of things rather than just describing them (which, strictly speaking, was what any sort of "natural history" was really about). A true, philosophical taxonomy of nature, as Ray too acknowledged, would represent the world as God had made it and as God perceived it; it would amount to understanding the world. The barriers to attaining such knowledge—the barriers to knowing whether or not one actually had attained it—were formidable, and most taxonomists found it best to ignore them and hope for the best.

The most important practitioner of taxonomy in natural history by the middle of the eighteenth century was the Swedish physician Carl Linnaeus (1707–78), whose own position on the philosophical status of his work was quite sophisticated. His work again reveals the interweaving of the natural-philosophical and the instrumental, or operational, threads that were coming to characterize modern science, and is the most significant example in the eighteenth century of the kind of intelligibility that was attributed to taxonomic, tabular organization. It was primarily due to the success of Linnaeus's taxonomic program that biologists today still speak of the hierarchical groups class, order, genus, and species (the modern category between order and genus, family, became standard a short time after Linnaeus). Linnaeus's attempts at classification implicated the same basic philosophical considerations as the ones discussed by John Ray some decades earlier. However, whereas Ray had, in effect, thrown up his hands at the possibility of creating classificatory categories that were justified by anything other than pragmatic utility, Linnaeus still held out the hope that a truly "natural" classification might some day be achieved.

The realization of this ambition would involve identifying classi-ficatory characters (the characters to be used as criteria in assigning species to genera, genera to families, and so forth) that would pro-duce natural groupings. But how were "natural" groupings to be rec-ognized? The general desideratum, not too different from Ray's, was to achieve groupings at each stage of the classification that con-tained members closely similar in most respects, not just similar ac-cording only to particular, privileged criteria. Ray had thought that such a goal was unlikely to be reached if a rigid set of classificatory criteria were used. Linnaeus, by contrast, dreamed of achieving just that: firm criteria that, as well as being universally applicable, also yielded what looked like a natural classification. Nonetheless, the difficulties of actually achieving this result were so great that Lin-naeus contented himself, in his great Latin taxonomic works, with presenting what he himself called "artificial" taxonomies, designed to ease the labors of the working naturalist. These used clear crite-ria without pretending to yield natural classifications.

A prominent aspect of Linnaeus's approach to taxonomy was that, artificial or natural, actual or ideal, any classificatory system in natural history was always going to be *static* (fig. 2.2). This feature, held in common with the views of Ray and of almost all other nat-uralists of the period, gave Linnaeus's world a permanence that lacked any meaningful dimension in time. Species had come into ex-istence at the Creation, and they remained the same down to the present day. Natural classification simply meant an organization of types that would correspond to God's plan; Ray had said that we could never know this plan, whereas Linnaeus thought it possible that we could, through diligent and comprehensive empirical work. He looked to the day when a comprehensive descriptive knowledge of plant species might be attained, thereby bringing the ultimate goal within reach. The achievement of this apparently "natural" classification would never guarantee with absolute certainty that the scheme was identical to God's own, but it would be a pretty con-vincing argument. The static quality of a natural classification was securely founded in the experience that species do not change over

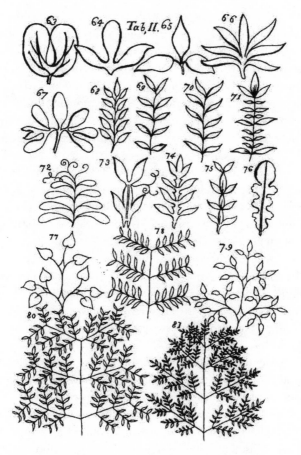

2.2. A page from an English translation of Linnaeus's Systema vegetabilium *("System of Vegetables"), 13th ed., showing types of compound leaves. The table relates distinct types by arranging them, nonetheless, according to relative similarities.*

generations; that like always breeds like, as farmers knew well. But it also reflected the divine, transcendental, and eternal character of God's knowledge of His Creation.

Linnaeus's conventionally pious understanding of the task of natural history was not universally shared among his contemporaries, however. The chief opponent of Linnaeus's kind of natural history was the great French zoologist Georges-Louis Leclerc, comte de Buffon (1707–88). In 1744, Buffon unleashed before the Royal Academy of Sciences in Paris an attack on Linnaean taxonomy, an attack that implicated all those involved in the taxonomic enterprise of which Linnaeus was the most prominent recent exponent. Buffon's central argument was essentially a version of Ray's from several decades earlier: Buffon held that the various hierarchically organized categories of species, genus, family, and so forth currently lacked any legitimate foundation in observation and clear thinking, and that any classificatory system relying on their use was therefore not acceptably philosophical. In other words, Buffon (an avowed admirer of Newton) wanted to justify natural history as part of natural *philosophy*.

The category of "species" was the only one that Buffon thought could be given an unequivocal and properly philosophical definition. In brief, his criterion for counting two individual animals as belonging to the same species was their membership in a common breeding community. This criterion, if it was met, overrode any other criteria based on the appearance of the animals; however different two individuals looked, if they could interbreed (directly or indirectly, depending upon their respective sexes), then they were of the same species, and that was that. For Buffon, the intelligibility of a taxonomic system was illusory; taxonomy could never be more than man-made, not simply because human beings were ultimately unable to tell when they had got it right, but because the very categories that it used were meaningless. In effect, Buffon held natural-historical classification to be unintelligible as a means of philosophical understanding. Mathematics, he argued, was also ultimately devoid of philosophical content, and on much the same grounds:

> In this century itself, where the Sciences seem to be carefully cultivated, I believe that it is easy to perceive that Philosophy is neglected, and perhaps more so than in any other century. The arts that people are pleased to call scientific have taken its place; the methods of calculus and geometry, those of botany and natural history, in a word formulas and dic-

tionaries preoccupy almost everyone. People imagine that they know more because of having increased the number of symbolic expressions and learned phrases, and pay no attention to the fact that all these arts are nothing but scaffolding for achieving science, and not science itself.[4]

The "arts" that Buffon mentions were those bodies of technique that were good for calculating and classifying but necessarily lacked insight into the causes and natures of things—the proper concerns of natural philosophy.

In fact, Buffon's practical natural history, represented by the many volumes of his great work *Histoire naturelle*, involved much detailed description of animals. The *Histoire naturelle* eschewed Linnaean-style taxonomy in favor of morphological description, accounts of characteristic behaviors and habitats, and even the animal's use to human beings. This last, apparently surprising element of Buffon's natural history reflected his belief in the importance of the senses and empiricism in learning about nature (a doctrine associated with Newton and the philosopher John Locke). Accordingly, human uses of animals and animal products represented the most intimate practical knowledge of those animals in human acquaintance with organic nature. Because of its stress on understanding the ways of life of animals in their environments, Buffon's approach could almost be labeled (anachronistically) as "ecological"; but it also tended to conceive of natural history as an element of man's dominion over nature—a particular form of instrumental knowledge that differed, in Buffon's own terms, from true natural philosophy.

The instrumental uses of taxonomic natural history were also manifold. Linnaeus's concern to develop taxonomic schemes for all three of the traditional kingdoms of nature—animal, vegetable, and mineral—sprang first from pragmatic considerations. Linnaeus's dream of a natural classification was just the philosophical counterpart of the effective instrumentality offered by his so-called artificial classifications (taxonomy seen as a handy tool for overworked naturalists); that usefulness resulted in the widespread adoption of his systems despite Buffon's philosophical cavils against it. Linnaeus's artificial taxonomy proved to be more valuable than Buffon's non-systematic descriptive approach: the instrumentality of Linnaeus's system shielded, so to speak, the natural philosophy of his hoped-for "natural" system, and kept the spark of the latter ambition alive.

III. Natural Classification

The prevalence of taxonomic systems during the eighteenth century resulted from the concerns of both natural philosophy and instrumentality, the intertwining pillars of modern science. But there was also another, especially powerful, incentive behind their development, which was their supposed pedagogical value. It had long been believed that an appropriate organization of material could facilitate its learning by the young, and a widely used technique during the sixteenth and seventeenth centuries had been a form of classification of subject-matters that strove to make memorization and recall easier to accomplish. The most popular scheme was associated with the work and theories of the mid-sixteenth-century French educational reformer Pierre de la Ramée, or Ramus (1515–72). It worked by taking a given subject-area, such as grammar (although it was applicable to anything), and proceeding to subdivide it dichotomously. Thus the first step was to divide the subject into two parts, each of those parts then being divided in two again, each of the resulting divisions again being susceptible to division into two parts, and so on for as far as one wished, or found useful.

Eighteenth-century classificatory ambitions therefore attached quite naturally to the structure of textbooks, regardless of the implied meanings of any particular taxonomy. But those meanings were of great significance from the standpoint of natural philosophy. Pedagogical ambitions relied on notions of how the mind most effectively learned, and the layout and presentation of material could therefore be as "artificial" as one liked so long as it was pedagogically effective. However, many of the taxonomies produced in this period for such diverse subjects as diseases, drugs, mathematical curves, chemical

species, and many other domains besides plants and animals purported, at some level or another, to capture something real about the
entities being classified. This uneasy expectation appears in Linnaeus's work, and the lesson may be applied in almost any area for
which taxonomies were proposed in the eighteenth century. Classification was not just cataloguing, the imposition of arbitrarily selected ordering principles; it was not mere "natural history" in the
usual sense of the straightforward description of nature's contents.
Eighteenth-century taxonomists generally aimed at being natural
philosophers who could provide the conceptual understanding that
would raise the status of their work and establish it as philosophically
meaningful, not simply as useful.

Indeed, many such taxonomies were *not* particularly useful; instrumentality often took second place to natural philosophy. Classification of minerals on the basis of crystal form, for example, did little
for geologists or mineralogists who wanted to locate useful associated
deposits of coal, for example. While knowledge of the crystal forms
of individual kinds of mineral might help the novice in making rapid
identifications, the idealistic taxonomic arrangement of such substances had no immediate or evident practical ramifications. But
it allowed such noted mineralogists as René-Just Haüy (1743–1822),
around 1800, to make what they saw as progress in understanding
mineral structure. Similarly, nosology, which typified and classified
diseases, was of little help to the practicing physician, but held significant philosophical content regarding *theories* of disease. Those
who regarded nosology as worthwhile regarded diseases themselves
as specific, distinct entities that the classification codified. By contrast, those medical theorists who regarded disease as a variable af-
fliction that expressed some kind of imbalance in the body saw no legitimacy to nosology, because they reckoned that it arbitrarily broke
up what was in reality a continuous spectrum of symptoms.

As long as natural philosophy remained an enterprise that held
a greater social status than did the artisanal ability to accomplish
material change in the world, the value of classification as part of
a knowledge-enterprise rested chiefly upon abstract, natural-

philosophical values. Thus Linnaeus's largely practical natural history nonetheless involved the explicit assumption that the ultimate philosophical quest was to find God's blueprint for nature. Buffon's assault on that taxonomic endeavor did not deny the place of God, but it did deny to taxonomy the ability to provide any kind of real understanding. For Buffon, the idea that human beings could discover a "natural" classification of organic nature that reproduced that of God, the Creator, made no sense; not only was it unachievable, but it was also unintelligible—the imagination was unable to grasp what such an achievement could possibly mean. Continuity of nature, rather than the atomization of it into discrete parts such as distinct species, made more sense to Buffon, just as it did to those medical theorists who distrusted nosology. But a fundamental natural-philosophical outlook that stressed continuity was on the defensive in this period; in any case, it was not well suited to competing successfully with classificatory schemes that offered pedagogical advantages in all subjects. Taxonomy worked well in the classroom, and colonized the minds of successive generations.[5]

But approaches were not uniform even within the confines of this taxonomic mentality. In the second half of the eighteenth century, there emerged two chief claimants to being the correct method of establishing true classifications in natural history. The dominant one was Linnaeus's explicitly artificial technique, in which a few delimited and preestablished criteria were employed to enable ready classification. In effect, Linnaeus's approach resembled a simple classification of books in a library. One can quite easily arrange the books based on an alphabetical arrangement of their authors' surnames, and Linnaeus took a roughly similar approach to classifying plants. Rather than privilege authors' names, he privileged particular properties of the plants' reproductive parts, the flowers and the fruit. Then, by means of counting such things as the numbers of stamens and pistils in the flowers, the Linnaean taxonomist would know, by definition, in which classificatory box to place a new kind, and could ignore other, irrelevant characteristics of the plant. Linnaeus justified his focus on reproductive parts by reference to Aristotle's teaching, which held that means of reproduction reflects central aspects of an organism's essential nature, but given the logic of his system he might as well have chosen any other characters whatsoever. That is why he referred to his as an artificial system of classification.

By contrast, an important French taxonomist, Michel Adanson (1727–1806), sought a "natural" system that he thought attainable simply by using a different technique. Spelled out in his book *Familles des plantes* ("Families of plants," 1763–64), this approach came at the problem from the opposite direction. Rather than relying on a small, privileged number of characters, in Linnaeus's fashion, Adanson determined to classify by differences rather than by similarities. His approach, briefly described, involved comparisons between species of plant so as to determine which of their features differed from one another. Similar plants would have few differences, but as types of plants grew more and more dissimilar, so an enumeration would display an increasing number of characters in which they differed. That increasing number then justified and measured the increasing separation between the two plants in the taxonomic layout. Adanson regarded his approach as productive of a natural classification because it did not require him to pick out a small number of characters, to act as taxonomic criteria, in advance. What he saw as Linnaeus's inability properly to justify his own selection of privileged characters was overcome, not by coming up with a convincing justification for just a few, but by employing *all* conceivable characters. Focusing on difference rather than similarity in turn avoided the need to specify every one among the multitude of possible characters; the naturalist need only mention those characters (usually only a handful) in which the plant under consideration differed from other apparently very similar ones.

Adanson's trick nonetheless involved the same fundamental perception of what a properly philosophical natural history should look like—that is, it appealed to the same intuitions regarding what kind of knowledge would or would not make sense of organic diversity. Like Linnaeus, Adanson treated taxonomic structures not as theoretical representations of *something else*, but as inherently intelligible demonstrations of the closeness of relationships between organisms. And closeness of relationship had nothing whatever to do with heredity, descent, or literal family relationships; relationship here was purely formal, much as one might say that a hexagon is more closely related to a pentagon than it is to a hyperbola—or a cornflake.

A major reconceptualization of taxonomy in natural history began to take place towards the end of the century, once again in France. Some naturalists began to conceive of new ways by which a taxonomic system, of which Linnaeus's had long provided the model for practical categorization, might be established as natural. Unprovable assertions that one's own scheme mirrored God's had always been problematic, as both Ray and Buffon had clearly recognized; even Adanson's attempts relied on intuitive plausibility rather than formal demonstration. But France, in the decades around 1800, fostered a scientific community whose highest values were those of mathematical precision and demonstration, as exemplified in the mathematical physics and scrupulous observational astronomy promoted by the great mathematician Pierre-Simon Laplace. Any mere naturalist who wanted his work to be seen as scientific, and his results truly contributions to natural philosophy, needed to strive for a system that possessed stern formal characteristics and rules of inference.

Antoine Laurent de Jussieu (1748–1836) worked at the state institution of natural history of which Buffon was, until his death in 1788, the head—the Jardin du Roi (or "King's garden," also called the Jardin des Plantes). Jussieu's great classificatory work, the *Genera plantarum secundum ordines naturales disposita* ("Kinds of plants arranged in natural order") was published in the revolutionary year 1789, and, as its title indicates, it pretended to a taxonomic system that was unequivocally natural. This book was to be regarded by later naturalists as pathbreaking, because they found Jussieu's claim to have developed a natural system of classification very persuasive. Naturalists, both botanists and zoologists, were evidently ready to accept such claims if they carried at least some philosophical plausibility.

Jussieu's philosophical weapon was something that he called "sub-ordination of characters," a concept that was shortly afterwards adapted to the demands of zoology by the leading French naturalist of the early nineteenth century, Georges Cuvier (1769–1832). Jussieu's idea of the "subordination of characters" was a way of expressing the hierarchy of classificatory criteria, used in assigning a species its place in a comprehensive taxonomic system, so that the hierarchy ap-peared to flow from a formal analysis of the structure of plants them-selves rather than from the arbitrary choices of the naturalist.

Jussieu generally picked the plant's mode of germination as the criterion for the broadest categories, representing this character as the most fundamental feature of a plant's function. The successively nested lower, and hence less fundamental, boxes of the categoriza-tion were in turn associated with characters that represented (ac-cording to Jussieu) ever-more subordinate properties relating to such things as root systems, leaves, and so on. The crucial feature of the entire scheme was that Jussieu could make a plausible argument, for each choice of classificatory criterion at each taxonomic level, that the chosen character was associated with a function less important than, and subordinated to, those above it, and at the same time more important than those below it.

Jussieu's work highlights a basic presupposition of "natural" clas-sification. His *Genera plantarum* contains a prefatory review written by fellow members of a leading French organization, the Royal Society of Medicine. This preface spells out the assumptions on which Jus-sieu's own work was based:

> M. de Jussieu's work has as its goal the gathering of all known plants [*Végétaux*] in an order that, interrupting none of the natural analogies by which different individuals of this kingdom appear linked together, presents them, on the contrary, in a sequence so continuous in nuances and relations [*rapports*] that this chain has no more need, in order to be complete, than the addition [*réunion*] of plants that naturalists have not yet discovered or observed: this is what he calls the natural method.[6]

The ideal that Jussieu's colleagues describe here is one in which the entire "vegetable kingdom" is imagined as a complete, uninterrupted fabric. The "continuous sequence" that they mention is not simply a conventional arrangement of plant descriptions in a book; it is supposed to represent the way that plants really are in nature. The continuity is one that Jussieu's method reveals, and that actually resides in the breathtaking multiplicity of organisms that populate the globe. Each plant in this "chain" of interlinked brethren will be only slightly different from the one preceding it, and any apparent gaps will surely be filled, in time, by the addition of plants "that naturalists have not yet discovered or observed." Unquestionably, the implication ran, such plants surely existed somewhere in the world.

This is a remarkable belief. It was shared by many other eighteenth-century naturalists, including Jussieu himself. It reveals a particular way of engaging with the world—and not just the organic world. At its root lies the assumption that everything that *can* exist *does* exist. The universe, God's Creation, is a panoply of realized possibilities, and the possibilities represented by terrestrial vegetable creation could therefore be expected to find actualization in real plants found somewhere on the earth. Plants were particular types that were snapshots of points along a continuum of creation, rather than different combinations of systematic taxonomic criteria of the kind found in Linnaeus's artificial system. The obvious implication of this view was that there should be no absolute line of division, or distinction, between one group of plant species and its nearest neighbor; each should blend, imperceptibly, into the next. This is how the paraphrase of Jussieu's ideas, written by the Royal Society of Medicine's reviewers, characterized the issue:

> [Jussieu's] method, supposing it to be complete, would not present, like artificial methods, great intersections, well-marked divisions; but each plant placed between its analogues would always find itself among family, and one could not say that it began or ended a particular series. Nevertheless, one would see in this great assemblage different modifications of the same organization forming, on the basis of proper analogies [*analo-*

gies spéciales], principal groups of very similar species; these groups placed each beside the others would touch each other and mix together by the undetectable [*insensibles*] nuances of their terminal species, always distinct in their centers, but always united in their extremities.[7]

As each group of related species blended into the next, there would still be no doubt that the groupings themselves had a real existence in nature, because comparison of the inhabitants of the core regions of the different groups would always show clearly their characteristic differences. Those differences simply cried out to be recognized; whether they were of different species in the same genus, or of different genera in the same family, they were not artificially conjured up by taxonomists. Through Jussieu's method, so the argument ran, order could be established in a world filled with fully realized possibilities. This view offered an understanding of organic nature that would be more than mere cataloguing, because its catalogue entries were not arbitrary. Instead, they revealed the ways in which the world itself was put together, which in turn showed how the differences between individuals were systematic and, in that sense, intelligible.

But Jussieu wished to offer his approach as more than just a matter of natural philosophy. In 1774 he had described his new method of organizing the plants at the Jardin du Roi:

> This order, which is that of nature, does not interest only the natural philosophers, but is of a more real usefulness. Rational analysis, confirmed by experience, shows that plants that agree in characters also have the same properties, so that once the natural order is given one could determine the properties by means of external marks.[8]

Natural philosophers, that is, are not interested in the (primarily) therapeutic uses of plants. They can come to understand those plants, however, through a genuinely natural classification; as a consequence, uses of the plants can be inferred in certain cases by the proximity, within a natural system of classification, of unknown plants to others that are already known to have the same desirable properties. Natural-philosophical understanding is therefore potentially of instrumental value, which is—according to Jussieu—a virtue that supplements, though it does not displace, those of natural philosophy itself.

IV. Reinterpreting the Natural

Classification of the kind practiced by Jussieu, Adanson, and other naturalists of the later eighteenth century also played its part in making sense of the heavens. The presumption that the universe was filled with actual realizations of its possibilities acted as a lens through which to view the world for the German expatriate William Herschel (1738–1822), the great astronomer and cosmologist (and not entirely insignificant musician) of late eighteenth- and early nineteenth-century England. One of Herschel's greatest achievements lay in his observational work on nebulae, those objects in the night sky that appear as small, faint cloudy regions (*nebula* is a Latin word for "cloud"). Using what were, for the time, large reflecting telescopes that he made himself, mirrors and all, Herschel spent much time in the 1780s and '90s observing, cataloguing, and describing many nebulae and star groupings. Some of this descriptive work involved a classificatory endeavor that resembled that of the naturalists: Herschel described it as a "natural history of the heavens." One of the issues that concerned Herschel was the true nature of nebulae, and what sorts there might be. How many, if any, were really regions of glowing cloudiness (perhaps gas of some kind), and how many were very distant stellar clusters that could not be resolved by current telescopes into the individual stars that composed them, but instead were seen only by virtue of their aggregate dull glow, much like the familiar Milky Way?

Herschel eventually developed, in the 1790s, a classification scheme that had at one end the so-called planetary nebulae, so named because their disklike appearance resembled that of a planet seen through a telescope. Herschel found that these nebulae some-times had a star at the center, and conjectured that they consisted of a gaseous sphere surrounding the star, and that the star had orig-inally formed through gravitational coalescence from a gas cloud. Given enough time, all the gas would eventually collapse into the star. After that, he suggested, the continued action of gravitation between the stars themselves would create loose groupings of stars, corre-sponding to the undoubted star clusters that Herschel and others had observed. These star clusters would become more densely packed over time, eventually becoming "globular clusters," a nebular vari-ety also identified by Herschel. Laid over this conjectural develop-mental process were crucial questions regarding the meaning of the observations that Herschel relied on: during observation of these var-ious sorts of nebulae, was the particular telescope used big enough, or the particular nebula close enough, to make visible the individual constituent stars that made up a dense star cluster at the end of the sequence? Only if those questions could be answered would it be possible to distinguish such a nebula from the cloud of glowing gas located at the beginning of the sequence.

Herschel's classificatory scheme mapped a continuous develop-ment over time, from gaseous cloud to globular cluster of stars, that possessed the same continuity as that of Jussieu's ideal "chain" of plant-groups that blended into one another. Herschel's stress on the taxonomic character of his scheme for objects in the heavens had the aim of identifying the kinds of nebulae to be found in the sky, much as the naturalist identified the kinds of plants found on the earth. And like the naturalists, Herschel believed that his types were *natural kinds*, despite their blending into one another in the course of their individual developmental histories. In a paper of 1791, Herschel compared himself to the natural philosopher who traces life down-wards from the most organized species of animal until, "when arriv-ing at the vegetable kingdom, he can scarcely point out to us the pre-cise boundary where the animal ceases and the plant begins. . . . But, recollecting himself, he compares, for instance, one of the human species to a tree and all doubt on the subject vanishes before him."[9]

Despite the dimension of temporal development, which served to link together the successive items in Herschel's continuous chain of celestial "species," his construal of the heavens represented them as a field containing certain specific kinds of objects, fixed in his taxonomic scheme even though not fixed in time. Indeed, the only real role for time in his cosmology was as one of the means (together with Newtonian gravity) by which his classification scheme could be certified as natural. Eighteenth-century taxonomy was never about long-term change; evolution arrived only in the nineteenth century, ironically in the wake of another non-evolutionist, the Frenchman Georges Cuvier.

Cuvier dominated the post-Revolutionary incarnation of the old Jardin du Roi, now known as the Museum of Natural History, from the mid-1790s until his death in 1832. He was throughout his life resolutely antievolutionary, even though his ideas and indefatigable research rescued taxonomy from the kind of doubts that naturalists, especially his predecessor Buffon, had tried to load on it during the previous century. Cuvier credited Jussieu with having originated the central conception that Cuvier brought to his own work on animals, namely "subordination of characters." But where Jussieu had continued to portray his botanical classificatory scheme as natural on the basis of its mapping onto the continuous chain of living beings, Cuvier rested his own claims to having developed a natural classification in zoology on *functional* considerations.

Jussieu had used a hierarchy of classificatory characters to structure his taxonomic techniques. He justified the hierarchy in generally functional terms to do with the relative importance of the characters and what they represented: he divided plants into three principal classes on the basis of features of their flowers (stamens, pistils, and so forth), on the grounds that these characters, conveniently constant among wide groups of plants, concerned organs essential for the plant's most fundamental function, that of reproduction. Cuvier, by contrast, set up the functions themselves as the items to be hierarchically arranged, only then identifying the characters that best expressed those functions. In this way he moved the spotlight away from classificatory techniques to the theoretical justification for those techniques, while at the same time representing Jussieu as a predecessor who shared his own understanding of what natural classification should mean.

In practice, of course, Cuvier's way of classifying animals owed a great deal to already-established procedures in zoology, including those of Linnaeus and of Cuvier's immediate French zoological predecessor, Félix Vicq d'Azyr (1748–94). The real novelty was Cuvier's insistence that his choice of classificatory criteria was entirely non-arbitrary. Every choice of a character as a criterion for classification had to be associated with the particular function that it served in maintaining the animal's viability. For example, teeth provided a standard criterion for distinguishing between different species within the same genus. Cuvier's earliest great achievement involved precisely this character, and derived its impact from a further implication concerning fossils and extinction.

At the end of the eighteenth century, fossil bones of mammals had periodically been found in Europe and in the Americas as well as in northern climes such as Siberia. Among them were bones and tusks resembling those of elephants. There was, however, a general reluctance to conclude that any of these remains represented vanished species. Extinction in nature seemed fundamentally counterintuitive in a world that was seen by most as well-structured and balanced; only the interference of man could exterminate an entire species, as in the celebrated case of the dodo in seventeenth-century Mauritius, wiped out by rapacious Dutch sailors. Cuvier managed to persuade naturalists otherwise.

The common assumption was that the fossilized bones were the remains of kinds of mammals still in existence. In the case of the elephants, it had been easy to say that the remains came from a variety of a still-living species, because there was no consensus on what it would take to count them, with certainty, as a wholly distinct species. But Cuvier's new arguments allowed him to overcome this presumption, with the aid of the collections at the Museum of Natural History as well as newly acquired ones shipped to Paris as a result of recent French military successes in the Netherlands.

In 1796 he argued, with the help of elephant skulls from Ceylon (now Sri Lanka) and southern Africa, that, contrary to earlier views, Indian and African elephants were taxonomically distinct species, not just different varieties:

> The first suspicions that there are more than one species came from a comparison of several molar teeth that were known to belong to elephants, and which showed considerable differences; some having their crown sculpted in a lozenge form, the others in the form of festooned ribbons.[10]

He went on to conclude: "It is clear that the elephant from Ceylon differs more from that of Africa than the horse from the ass or the goat from the sheep."[11] In the 1799 version of this argument, Cuvier stated his conclusion quite baldly: "no naturalist can doubt that there are two quite distinct species of elephants."[12] Once this point was established, the fossil remains became much more amenable to unequivocal classification.

> The teeth and jaws of the mammoth do not exactly resemble those of the elephant. . . . These animals thus differ from the elephant as much as, or more than, the dog differs from the jackal and the hyena. Since the dog tolerates the cold of the north, while the other two only live in the south, it could be the same with these animals, of which only the fossil remains are known.[13]

In the late 1790s, Cuvier strengthened his claims by providing explicit arguments to establish the legitimacy of using such characters as criteria for identifying distinct species of animal. In the published form of his lectures at the museum, the *Lessons in Comparative Anatomy* (1800–1805), Cuvier explained that classificatory distinctions of this sort were justified (philosophically, we might say) by the hierarchical

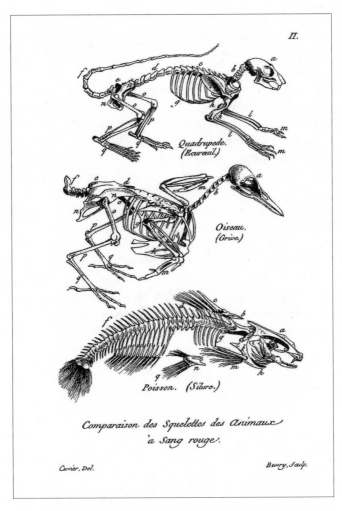

2.3. Cuvier's Tableau élémentaire de l'histoire des animaux *(French republican year 6, i.e. 1797/8), showing the skeletons of three very different sorts of animals (squirrel, thrush, catfish). Cuvier's "comparative anatomy" juxtaposed even widely different sorts of animals to emphasize how systematic differences between corresponding parts revealed the parts' suitedness to their very diverse functions (e.g. leg of squirrel and wing of bird). Darwin was to interpret these matters very differently, but still while valuing taxonomic comparison.*

functions that they represented. At the relatively fine level of discrimination necessary to distinguish different species of the same genus, teeth were appropriate criteria because they relate directly to the "nutritive function"—they reflect the importance of eating.

Cuvier made claims of broad natural-philosophical significance on the basis of classification. The arrangement of species within a taxonomic system held implications for extinction and the course of earth history as a whole. And the reason classification could be elevated from a descriptive, cataloguing procedure to a procedure that possessed natural-philosophical content was not, as it had been for many of his predecessors, because it represented God's plan of organic Creation, or because it provided a map of realized possibilities for living things. It was natural philosophy because it now represented the basic functional structures of animals, on the basis of which they were enabled to live in the world—to exist (fig. 2.3). In the France that followed the turmoil of the Revolution, the possibility of living successfully in the world had become something worth investigating, and worth understanding. The meaning of hierarchical classification no longer rested on disembodied ideas.

CHAPTER THREE

The Chemical Revolution
Thwarted by Atoms

I. Chemistry as Natural Philosophy

Natural-history tables were not the only way for a natural philosopher to understand chemistry in the eighteenth century. Indeed, a
central concern in that period was making chemistry into the right
kind of study for a natural philosopher in the first place. Chemistry
was an endeavor that was traditionally associated with practical
techniques—such as making dyes, tanning, preparing drugs and
ointments—or with alchemy. But by the beginning of the eighteenth
century, chemical phenomena were well on their way to becoming
integral parts of natural philosophy, not least because of their perceived role in learning about the underlying structure of matter.
Isaac Newton had advocated chemistry as a source of insight into
matter theory in his book *Opticks,* and a number of seventeenth-
century chemists had also adopted, like Newton, talk of corpuscles
or atoms as ways of accounting for chemical phenomena. When
Geoffroy, as we saw in the previous chapter, introduced his table of
rapports to the Academy of Sciences in Paris, he was attempting to
raise chemistry as a whole to a more respectable intellectual status
than simply that of a messy, overly complicated set of practical techniques, and to bring to chemical phenomena a clarity and order that
would totally displace the arcane, obscure, and spiritually engaged

purification rituals of alchemy. If chemistry were to become an integral part of natural philosophy, it needed concepts and procedures suited to such criteria of intelligibility as had been variously promoted by Descartes, Boyle, Newton, and others. At the same time, it needed to be teachable.

That was one of the reasons for the success of chemical tables in the eighteenth century. Textbooks of chemistry had begun to appear in the seventeenth century, heralded by a work of 1597 written by the German schoolmaster Andreas Libavius (1540–1616). These books detailed the apparatus and practical techniques of the chemist. However, they did not focus their efforts on establishing chemistry as a form of natural philosophy. When, by the end of that century, some natural philosophers had attempted to integrate chemical knowledge into their accounts of matter, the ambitions of chemists began to rise. But their problems were considerable. The distance between actual chemical processes and the conjectural explanations that some people provided for them, presented in terms of tiny, invisible particles, was enormous. If such explanations were all that a chemist could provide to support his claims to being a natural philosopher, he could never be much more than a speculator. Chemical tables, of the sort examined in chapter 2, provided one promising alternative.

They were not, however, the only one. Georg Ernst Stahl (1660–1734), a German physician, was a successful writer of chemistry textbooks in the first half of the eighteenth century. His approach combined a sophisticated natural-philosophical awareness of chemistry's tenuous intellectual status with a chemist's commitment to practical chemical procedures and the properties of chemical substances. Stahl's most influential book, *Fundamentals of Chemistry*, first appeared in 1723 and was quickly translated into several other languages, including English. The book begins with a presentation of the kind of speculative matter theory made familiar to natural philosophers by the writings of Robert Boyle and Isaac Newton. This theory postulated small particles as the underlying constituents of matter, with the ultimate particles being all alike. The differences between different kinds of substance was due to the various ways in which these basic particles were arranged to form more complex aggregate particles. Those aggregate particles could in turn be arranged together to form even more complex particles, and so on up an increasingly elaborate hierarchy of material constitutions. The variety of physical and chemical properties displayed by real substances was held, on this view, to result from the huge variety of possible particulate combinations.

However, after explaining this view, Stahl proceeded to discard it. He did not reject it as false; indeed, his discussion was intended to represent it as the most probable theory of material substances. But Stahl held that the chemist needed to use practically accessible, properly chemical concepts that would engage directly with actual chemical experience. Discussion of corpuscles was all very well, but it was of little help in the business of doing practical chemical work. While chemical phenomena, Stahl said, would perhaps one day be understood in terms of the arrangements of physical corpuscles, that day was still far off:

> as this is not easily obtainable from the Chemistry of these days, and so can hardly be come at by Art; a difference, at present, prevails between the *physical* and *chemical Principles* of mix'd Bodies. . . . Those are called *physical Principles* whereof a Mixt [i.e. a substance made up of several different basic components] is really composed; but they are not hitherto settled. . . . And those are usually term'd *chemical Principles,* into which all Bodies are found reducible by the chemical operations hitherto known.[1]

Stahl's position, then, was that although all chemical substances might ultimately be composed of the same kind of particles ("physical principles") in different arrangements, that belief was of no help in understanding chemical phenomena. These had to be understood in terms of "chemical principles." The latter term referred to chemical raw materials; real, accessible material things, familiar from laboratory experience, from which more complex substances appeared to be composed.

In the previous century, most chemists had believed that there was just a handful of basic elements making up all the diverse substances met with in the chemist's laboratory, usually Aristotle's four (earth, air, fire, and water) or the alchemist's three "principles" (salt, sulfur, and mercury). But following Stahl's lead, in the eighteenth century these kinds of elements steadily gave way to concepts that were designed to make sense of the chemical properties possessed by actual laboratory materials (many of them artificial productions provided from the apothecary's shop).

In the wake of Stahl's call for the scientific autonomy of chemistry, the 1740s and '50s saw a renewed interest in chemical tables. These tables codified taxonomically the reactive properties of many familiar chemical substances, and treated those substances themselves as the theoretical focus of attention in chemistry rather than undetectable fundamental particles. Chemical substances were things that the chemist could see, smell, taste, and manipulate; the widespread view that natural philosophy should be rooted in sensory experience therefore found a receptive audience among chemists in the second half of the eighteenth century.

II. French Chemical Experience

France was the leading country for chemical researches in this period. In the second half of the century, French chemists could choose between two major views on how experience should operate in the making of chemical knowledge. One of them was represented by Gabriel-François Venel (1723–75), a chemist who, like the great Antoine-Laurent Lavoisier (1743–94), had been a student of Rouelle's in Paris. Venel wrote several articles on chemical subjects for one of the greatest works of eighteenth-century publishing—the huge, multivolume *Encyclopédie*, edited by the mathematician Jean d'Alembert (1717–83) and the philosopher Denis Diderot (1713–84), which began appearing in 1751. Although the *Encyclopédie* tends to be remembered for its celebration of Newton, Venel's article on chemistry explicitly associated it with an approach to nature that was less formalistic and less mathematical than the stereotypical Newtonian approach to natural philosophy.

Venel praised as his ideal kind of chemists those rather mystical practitioners (we would call them alchemists) who had preceded Newton. Venel thought that they had seen more deeply into nature than just the descriptive surface; in a way, his sentiments were similar to those of Newton's critics at the end of the seventeenth century, people like Huygens, who distrusted Newton's work in the *Principia* because it failed to get at the causal roots of phenomena. Chemical predecessors of Newton's saw more of what nature was in itself, according to Venel: "a sympathy, a correspondence, was for them just a phenomenon, whereas for us, as soon as we cannot relate it to our pretended laws of motion, it's a paradox."[2] Venel advocated a kind of intelligibility different from that favored by mechanists. Stressing firsthand experience of the natural world, he made the phenomena themselves the foundation of normality: when the practical chemist found that two substances seem to have an affinity, a "correspondence" for one another, there was nothing to be explained; it was simply a phenomenon, and could be understood as such. But for the Newtonian natural philosopher, who wanted to make sense of the world by reducing all phenomena to a limited set of fundamental principles, chemical behaviors posed problems that seemed to be intractable. How indeed, Venel wondered, could chemistry be explained by Newton's laws of motion?

At the start of the nineteenth century another French chemist, Claude-Louis Berthollet (1748–1822), attempted to answer Venel's question. Berthollet wanted to interpret the familiar affinity tables in terms of short-range forces acting between the elementary particles of different chemical substances. Differential affinities would be explained by differences in attractive forces between particles, which in turn were explained by postulated differences in the particles' shapes. Berthollet went so far as to suggest that these forces might be identical to gravitational forces: although a single particle of a chemical species would exert only a minuscule gravitational attraction, the particles attracted to it could get very close to its center if it were very dense. As a result, the inverse-square attractive force might be able to grow enormously as the particles drew nearer to one another. But despite this theoretical possibility, Berthollet was unable to get his scheme to work in any detail, and eventually abandoned it in favor of more tractable problems.

Venel's remarks had taken the existence of such difficulties for granted; he had rejected the whole Newtonian approach as one that had to resort to labeling any inexplicable phenomenon a paradox. For him, the lesson was clear: the practical chemist understood the behaviors of substances almost instinctively, through intimate familiarity with them. It was an understanding that could not always be adequately verbalized, in the same way, perhaps, as a good cook's skills cannot be wholly captured in a written recipe. That, for Venel, was the essence of being a chemist; it was how he engaged with chemical behaviors and properties. Chemistry was a practice as much as it was a body of abstract theoretical knowledge, and as such it became intelligible.

The other major understanding of how experience could build chemical knowledge is exemplified above all by Lavoisier; his kind of chemistry was to become dominant by the nineteenth century. Lavoisier proclaimed a "revolution" in chemistry in the 1770s, a revolution that a host of his followers carried through successfully. One of its most touted features was the use of quantitative experimental data, an approach that Venel had condemned as yielding only superficial knowledge. Lavoisier championed a way of doing chemistry that aimed to treat it as just another physical science. It would be most successful when it was managed by means similar to the mathematical techniques of the astronomer or the theoretical mechanician—the techniques associated with the great name of Newton.

Experience was central to Lavoisier's arguments for his new chemistry, and he was perfectly explicit about how to understand and use it. He took his arguments from a contemporary, the French philosopher Étienne Bonnet, abbé de Condillac (1714–80). Condillac saw himself as following in the footsteps of Newton's admirer John Locke, who had investigated the ways in which human understanding of the world is constructed by the mind from the sensory data that it receives. Condillac wanted to explore this theme in a methodical way; he did so above all in his *Traité des sensations* ("Treatise on sensations") of 1754, in which he developed an account of how a self-consistent picture of the external world might be formed in the mind of a man. He imagined an elaborate statue of a man, possessing all the internal organization of a real man but lacking any ideas in its mind as well as any bodily senses. Then Condillac imagined this statue acquiring, one by one, the five senses, starting with that of smell and ending with that of vision. By this means, Condillac intended to display the precise ways in which elaborate concepts could be formed on the basis of nothing but the senses and the mind's reflection on their lessons, showing what could be learned from each individual sense as well as how each sense coordinated with the others. He wanted to show that human beings can create a self-consistent mental representation of the physical world without needing any inborn ideas to give them initial traction—a rejection, in other words, of the so-called rationalist view held by the Cartesians, which argued that knowledge of basic truths (such as those of formal logic and mathematics) could be known by the reasoning mind itself, without need of the senses.

Condillac's Lockean empiricism was favored by most liberal philosophers in eighteenth-century France, including Voltaire, and it was adopted wholeheartedly by Lavoisier. It is always difficult, if not impossible, to say whether any writer's professed philosophical allegiances are "real" or merely adopted for their presumed rhetorical effect; certainly, Lavoisier chose well by presenting his chemical theories as if they were necessarily undergirded by the fashionable philosophy of Condillac. Lavoisier's preface to his greatest work, the *Traité élémentaire de chimie* of 1789 (usually known in English as his *Elements of Chemistry*), used Condillac as the central justification for a form of empiricism that ran through all of Lavoisier's mature chemical work: it was the way that Lavoisier made sense out of chemical experience.

Early on in his preface, Lavoisier remarks, like a scientific Jane Austen:

> It is a maxim universally admitted in geometry, and indeed in every branch of knowledge, that, in the progress of investigation, we should proceed from known facts to what is unknown. In early infancy, our ideas spring from our wants; the sensation of want excites the idea of the object by which it is to be gratified. In this manner, from a series of sensations, observations, and analyses, a successive train of ideas arises, so linked together, that an attentive observer may trace back to a certain point the order and connection of the whole sum of human knowledge.[3]

Lavoisier goes on to explain the analogy found in the sciences: "in the same manner, in commencing the study of a physical science, we ought to form no idea but what is a necessary consequence, and immediate effect, of an experiment or observation."[4] Alas, he continues, the negative consequences that flow from neglecting this rule are much less severe than those attending a child that neglects its physical welfare: the perils of imagination and pride have often led us astray, with old assumptions, sanctioned by authority, remaining unquestioned. Lavoisier then reveals that, to avoid going astray, he had submitted himself to a law whereby he would never "form any conclusion which is not an immediate consequence necessarily flowing from observation and experiment."[5] This determination, he admits, leaves his book with a surprising gap in its intended systematic coverage of chemistry:

> The rigorous law from which I have never deviated, of forming no conclusions which are not fully warranted by experiment, and of never supplying the absence of facts, has prevented me from comprehending in this work the branch of chemistry which treats of affinities, although it is perhaps the best calculated of any part of chemistry for being reduced into a completely systematic body.[6]

In other words, Lavoisier elects to leave aside elective affinities and their famous tables, because they go too far ahead of direct experience.

Lavoisier's announced intention amounted to claiming that he would lay out the solid foundation upon which true chemical ideas (his own) were based, and on which the future course of chemistry should be developed. He ends his preface by quoting at some length from Condillac:

> At the end of the fifth chapter, the Abbé de Condillac adds: "But, after all, the sciences have made progress, because philosophers have applied themselves with more attention to observe, and have communicated to their language that precision and accuracy which they have employed in their observations: In correcting their language they reason better."[7]

Language was central to Lavoisier's project; two years earlier he and three colleagues had published a systematic reform of the language of chemistry that they claimed was based directly on experiment and observation. The simplest kinds of chemical substances were to be given simple names, and chemical compounds would have more complex names that indicated the simpler ones from which they were compounded—using as the basic naming technique a binomial nomenclature modeled on Linnaeus.

Like Linnaeus's names for animal and plant species, Lavoisier designated basic compounds by a generic name plus a specific name. Thus, for example, the genus "sulfates" contained species distinguished from one another by the fact that different substances were involved in otherwise similar reactions: hence sulfate of iron, sulfate of nickel, sulfate of copper, and so forth.[8] The naming of these compound substances was intended to reveal their means of formation, or creation, by the chemist: these were all substances that were formed by the combination of sulfuric acid with various "salifiable bases," that is, substances characterized by their capacity to form salts (such as sulfates) when reacted with sulfuric acid. Lavoisier's names for chemical substances were intended to summarize the practical chemical experience of making them or of using them.

Correspondingly, Lavoisier needed to give single names to those substances that appeared to be absolutely simple; that is, substances that seemed not to be compounded from two or more simpler ingredients. Simplicity was a strictly practical issue, in that the only criterion for treating a substance as simple was that chemists had so far found no way of decomposing it into constituent parts. This accumulated (negative) experience yielded a list of what Lavoisier called "simple substances . . . which may be considered as the elements of bodies."[9] These substances received single, uncompounded names, reflecting their apparently uncompounded natures. Lavoisier's chemical alphabet was a sophisticated response to Stahl's call for properly chemical concepts that would abandon the quest for the true physical constituents of substances; Lavoisier aimed to define chemistry in terms of laboratory pragmatism.

The intelligibility of Lavoisier's chemical natural philosophy rested on accepting a huge amount of practical chemical experience not all of which any individual chemist would have been able to acquire personally. Furthermore, it was a natural philosophy that refused to probe to the farthest depths of the natural world; it could not speak of the ultimate constituents of the material universe, but only of what the world of experience seemed to contain. It was a natural philosophy that was not so much a philosophy of nature as a philosophy of *inquiry* into nature. As such, it earned a good few enemies.

III. British Sense versus French Nonsense

One of the most vehement opponents of Lavoisier's chemistry and its new nomenclature was the English (and, eventually, American) natural philosopher Joseph Priestley (1733–1804). Priestley, together with other British chemists, objected to what they saw as Lavoisier's dogmatic presentation of his work and ideas. They asserted that, contrary to Lavoisier's own claims, his chemistry went well beyond what was warranted by experiment and observation. Priestley was, and remained until his death in 1804, the foremost champion of an idea earlier popularized by Stahl, the theory of phlogiston, that Lavoisier's chemistry rejected.

Phlogiston was supposed to be a substance that explained the phenomena of ordinary combustion. A body that could be set on fire was, on this view, a body that contained phlogiston; that fact explained why it was combustible. When the body burned, it was in fact releasing phlogiston into the air, as anyone can see by its rising flames. The loss of phlogiston left behind the dead shell of the original matter, no longer plump with its active phlogiston. The most chemically controllable examples of combustion were those of metals, a process known as calcination. Unlike, say, burning wood, where much smoke carries off most of the combustion products, the calcination of metals left behind a good quantity of dull, lumpen material that was known as the metal's *calx* (hence the term "calcination"). Phlogiston chemistry could account for the bright, shiny appearance of the original metal in contrast to the dull appearance of its calx by attributing the former precisely to the presence in the metal of its phlogiston: once the phlogiston had been driven off, the residue no longer displayed those characteristics associated with metals.

Lavoisier's chemistry made sense of the same phenomena in a quite different way—indeed, a way so different that the features of combustion that were seen as important were themselves different. Although Lavoisier, like Priestley, spoke of calcination and combustion, for him their most essential characteristic was not the flame, but the change in weight of the combustion products as compared with the weight of the original material. This feature was most clearly displayed in the cases of metallic calcination: burn the metal in a closed container, so as to avoid loss of the combustion products, and the weight of the calx typically exceeded that of the original metal. According to Lavoisier's reported experimental results, some of the air in the container had also been used up in the course of the combustion. Such experiments and reasoning led to his assertion that combustion involved a chemical process in which a particular component of the atmosphere (called, in the new French chemical nomenclature, *oxygène*) combined with the metal to yield a chemically distinct substance, the calx, which was therefore now seen as a chemical combination of oxygen and metal. Where for Stahl, Priestley, and most other chemists, the metallic calx had been chemically *simpler* than the metal, now it was more complex, and the metal was the more basic substance.

The criterion of weight was central to Lavoisier's argument in a way that was not typically the case for other chemists, those who valued more highly the qualitative features of chemical experience. Lavoisier's new chemistry stressed the demonstrative importance of precise *quantitative* experiment in reaching chemical conclusions. Arguments based on quantitative results appeared to possess the accepted clarity and intelligibility of arithmetic. But Lavoisier, together with his French colleagues and supporters, was seen by English opponents to be placing a thoroughly unreasonable stress on quantitative precision, and using that precision as a means of forcing other chemists into silence. Lavoisier seemed to them to be saying (and sometimes did say) that if any would-be critic had not performed the elaborate experiments used to support his claims—and to the level of precision that Lavoisier had achieved—then that critic was not equipped to challenge him. This was an assertion that Priestley, among others, indignantly rejected. For Priestley, as for Lavoisier, properly scientific results in natural philosophy were based on experience. But Priestley did not think that experience in natural philosophy was the preserve of a privileged group of specialists who dictated to everyone else. Instead, natural philosophy should be established on experience that was accessible to anyone who was interested; the authority of natural philosophy ought not to derive from the say-so of an élite (still less an autocratic French élite), but from experiences available to everyone. Priestley complained that Lavoisier's experimental chemistry "requires so difficult and expensive an apparatus, and so many precautions in the use of it, that the frequent repetition of the experiment cannot be expected."[10] Lavoisier's often complex experimental setups, with their enormously

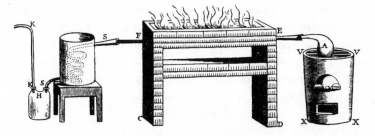

3.1. Lavoisier's experiment to analyze water (boiling in flask A) into its constituents of oxygen and hydrogen by passing the steam through a red-hot glass tube (EF) that contains spirals of beaten iron. The oxygen from the steam combines chemically with the glowing iron, while the released gaseous hydrogen is collected at left; scrupulous weighing verifies the equivalence of water and products. From Lavoisier's Traité élémentaire *(1789).*

high demands on quantitative precision, were far from being generally available.

Priestley was not the only chemist in Britain who criticized Lavoisier's chemistry for doing exactly what Lavoisier meant it to do—to impress readers by sheer weight of technical accomplishment. One of Lavoisier's most famous experimental arguments concerned his claim that ordinary water, which had traditionally been regarded as a simple, uncompounded substance, was in reality an oxide of a novel gas that was soon called, in the new nomenclature, *hydrogène.* Combustion of hydrogen generated water; hence the name (from the Greek *hydōr,* meaning "water"). But Lavoisier's methodological rules for chemistry required him not only to make water by burning hydrogen—the synthetic stage, as he called it—but also to analyze water back into its supposed constituents. He did this by conducting an experiment, in 1785, that required weight measurements of unprecedented accuracy and precision (see fig. 3.1 for a later variant).

Lavoisier's apparatus allowed the trickling of a small amount of (carefully weighed) water down the inside of an inclined gun barrel made of iron. The barrel was heated in advance to a red heat, after having itself been weighed. The idea behind the experiment was that the water would be split by the enormous heat of the metal into its constituent parts, oxygen and hydrogen. When this happened, the oxygen would combine chemically with the iron of the barrel, while the hydrogen evolved could be collected at the barrel's farther end. Confirmation of the transformation would be achieved by measuring the subsequent weight of the barrel, to show that its increase, due to the new rust, exactly equaled the proportionate weight of oxygen believed to have been contained in the water.

Lavoisier was very proud of the exactitude of his experiment.

> This double experiment [i.e., the analysis and a subsequent, equally quantitative synthesis], one of the most memorable which was ever made, on account of the scrupulous exactness which was attended to, may be regarded as a demonstration of the possibility of decomposing and recomposing water, and of its resolution into two principles, oxigene and hydrogene [*sic*], if in any case the word Demonstration may be employed in natural philosophy and chemistry.

He continued with a challenge to possible opponents of his conclusions: "it is by experiments of the same order, that is to say by demonstrative experiments which they ought to be attacked."[11]

Lavoisier treated quantitative reasoning as peculiarly intelligible. For him, the demonstrative character of reasoning that used quantitative measurements was increased if the *precision* of those measurements could be seen as very high; he habitually presented his experimental measurements to an unusually large number of significant figures, regardless of the measurements' margins of error. If this was intended to persuade Lavoisier's readers of the care with which he had done his work, it prompted just the opposite response among critics in Britain. One, the natural philosopher William Nicholson (1753–1815), echoed Priestley in speaking of Lavoisier's "unwarrantable pretension to accuracy," and went on to impugn the rhetorical effect of Lavoisier's numbers:

> If it be denied that these results are pretended in the last figures, I must beg leave to observe, that these long rows of figures, which in some in-

stances extend to a thousand times the nicety of experiment, serve only to exhibit a parade which true science has no need of: and, more than this, that when the real degree of accuracy in experiments is thus hidden from our contemplation, we are somewhat disposed to doubt whether the *exactitude scrupuleuse* of the experiments be indeed such as to render the proofs *de l'ordre demonstratif.*[12]

Nicholson scorned Lavoisier's assertions, and his words show how Lavoisier's tactics could backfire badly. Lavoisier wanted precise numbers from scrupulously conducted experiments to create a balance sheet. This balance sheet would carry not only conviction but also evident clarity and intelligibility to his readers—Lavoisier collected taxes for the government, so fiscal techniques made especial sense to him. Unfortunately, his chemical readers in what Napoleon was to call the "nation of shopkeepers" were less prepared than many of his French colleagues to swallow Lavoisier's accounting practices.

Lavoisier wanted his chemical science to refer only to things that had directly observable, preferably measurable, existence in the laboratory. But his approach had little appeal to those who wanted chemistry to contribute to a broader natural philosophy, one that spoke about the way nature truly was in itself. Lavoisier's weights and measures seemed to represent a rational system for assessing everything, but they failed to explain why things were as they were— much as the cynic is defined as one who knows the price of everything but the value of nothing. Lavoisier's chemistry was shallow, on this view, and its evident usefulness dodged important issues of natural philosophy. Priestley's narratives of his own experiments told stories of discovery instead of presenting pseudomathematical demonstrations; he wanted experience to lead the mind to what were often unexpected discoveries, not to demonstrate conclusions through the synthesis and analysis beloved of Lavoisier.

In his celebrated accounts of work on the air from the mid-1770s, work that often appears in standard textbook treatments of the discovery of oxygen, Priestley stressed again and again the surprise attending the outcomes of his experiments. He emphasized the importance of serendipity in "philosophical investigations," and declared that "more is owing to what we call *chance*, that is, philosophically speaking, to the observation of *events arising from unknown causes*, than to any proper *design*, or pre-conceived *theory* in this business."[13] He proceeded to stress the unexpectedness of the discoveries that he was about to relate, and how reluctant he had at first been to allow them as true.

> And yet, when I re-consider the matter, and compare my last discoveries relating to the constitution of the atmosphere with the first, I see the closest and the easiest connexion in the world between them, so as to wonder that I should not have been led immediately from the one to the other. That this was not the case, I attribute to the force of prejudice, which, unknown to ourselves, biasses not only our *judgments*, properly so called, but even the perceptions of our senses.[14]

Priestley then described his experimental path, which had led him through all manner of things that he was "utterly at a loss how to account for," or that occasioned his "surprize," or that were "remarkable"; how other things made him "much more surprized," and "puzzled."[15]

> I wish my reader be not quite tired with the frequent repetition of the word *surprize*, and others of similar import; but I must go on in that style a little longer. For the next day I was more surprized than ever I had been before.[16]

Strikingly, Priestley also characterized himself as *not* being "anything of a practical chymist,"[17] his expectations being less attuned to his materials than other people's might be. He wanted to be seen strictly as a natural philosopher rather than a practical chemist, and one primarily interested, in his case, in the atmosphere's constitution, its production (he guessed from volcanoes), and its relationship to vegetation.

Rather like Venel, Priestley approached natural philosophy as an endeavor in which the inquirer cooperates with nature; nature's intelligibility appeared in conversation with her—nature was (and still is) usually personified as feminine. Priestley presented his work as a kind of adventure, the final outcome of which was unknown until the end came in sight. By contrast, Lavoisier, in appealing to Condillac's philosophy, wished to represent his work as systematic, derived from the proper, disciplined interpretation of precise experience. In effect, he approached questions in chemistry as engineering problems that needed to be solved in as direct and straightforward a manner as possible. Lavoisier's critics disliked what they saw as his intellectual bullying, and a spurious exactitude that smacked of hubris. But Lavoisier's self-announced "revolution in chemistry" was grounded in a view of nature that took her as a logical puzzle, one that could be mapped out in advance with the help of a properly constructed language—a blueprint for how to understand the world.

IV. John Dalton and the Chemical Atom

John Dalton (1766–1844) was not a chemist. From 1793 onwards, he taught "natural philosophy and mathematics" at New College, Manchester, in the industrial north of England, and the particular focus of the work that produced his form of chemical atomism was, to begin with, meteorology. In the years around 1800, he had become interested in various problems concerning the behavior of the atmosphere, such as the relationship between heat and the expansion of gases, and the amount of water vapor that gases could hold at different temperatures and pressures—questions with obvious connections to the weather (especially that of notoriously rainy Manchester). He did a bit of experimental work on these things, especially having to do with the water-holding capacity of gases, but his real interest lay in understanding what, at an underlying natural-philosophical level, was really going on physically. Dalton's ideas about the gases in the atmosphere took the form of considering their supposed constituent particles.

The idea that the atmosphere is composed of a mixture of various distinct gases was then quite novel, having been established through the work of such people as Priestley and, above all, Lavoisier. But Dalton's approach to making sense of the behavior of these gases differed from Lavoisier's chemistry in crucial ways. Lavoisier wanted to speak of the behavior of gases, like that of all chemically distinguishable kinds of matter, strictly on the basis of their phenomenological properties as found in the laboratory. Dalton wanted to speak about them on the basis of what gases really were in themselves—a traditional concern of the natural philosopher. Dalton simply assumed that gases were really composed of minuscule particles that exerted repulsive forces on one another. That was the standard picture in Britain, and had been ever since the time of Newton: Newton had put forward the idea in his *Principia,* as well as in the *Opticks,* as a way to explain the elastic properties of the air. The repelling distance force increases as the distance between the air particles decreases, so that when a volume of air is compressed, its particles are forced closer together and their mutual repulsions increase; hence the compressed air tends to push back outwards. This conception had been adopted by Newtonian natural philosophers in Britain and elsewhere throughout most of the eighteenth century; Dalton used it as the obvious model, but applied it to the distinct gases of the atmosphere rather than simply to a generic "air" as Newton had done.

Dalton's attempt to understand the atmosphere focused initially on an apparent problem that arose from seeing it as made up of several distinct gases. Lavoisier had argued that air was primarily composed of two gases, one known at the time as "azote," or nitrogen, the other oxygen, in proportions of roughly 80 percent to 20 percent by volume. (The name "azote" was Lavoisier's invention, chosen to express, from Greek roots, its principal property of *not* supporting life.) Dalton thought that a mixture of different gases immediately created a physical problem. If a gas consisted of what he called an "elastic fluid" of mutually repelling particles, the question arose as to whether the particles of two different gases repelled each other—

whether an oxygen particle repelled a nitrogen particle as well as other oxygens, and so on. If that was the case, Dalton was unable to conceive why the heavier gas would not simply settle out, the lighter gas rising above it, just as occurred with immiscible liquids of different densities.

Dalton's solution, in a paper of 1801, was to suppose that the two principal kinds of gas in the atmosphere, and presumably different kinds of gas generally, should be understood as existing completely independently of one other. Oxygen particles repelled oxygen particles, and nitrogen particles repelled nitrogen particles, but between two different particles there were no forces of any kind at all; each was, so to speak, totally unaware of the presence of the other. In a gas mixture, then, every gas acts as an independent entity—an idea related to what came to be known subsequently as "Dalton's law of partial pressures." The two chief gases of the atmosphere did not settle out because they simply did not weigh upon each other.

Dalton continued to work with this picture as he considered further the solubility of gases in water. In the course of investigating its proportion in the atmosphere, Dalton noted that carbonic acid gas (nowadays called carbon dioxide) was particularly soluble. In a paper of 1802 on the subject, he wrote that carbonic acid gas "is held in water, not by chemical affinity, but merely by the pressure of the gas."[18] The gas particles were kept in solution by the pressure of that same gas on the water's surface. In words reminiscent of Priestley, Dalton wrote that the experimental results supporting his explanation were a "matter of surprise to me."[19]

Later that year, when the chemist William Henry (1774–1836) set out to show that, contra Dalton, gaseous solubility should be understood as a chemical phenomenon involving affinities, Henry found that Dalton's "physical" interpretation worked better. Dalton was pleased by this development, and wrote of it in a revealing way:

Upon due consideration of these phenomena, Dr Henry became convinced, that there was no system of elastic fluids which gave so simple, easy and intelligible a solution of them, as the one I adopt, namely, that each gas in any mixture exercises a distinct pressure, which continues the same if the other gases are withdrawn.[20]

Dalton's concern with this particular kind of "intelligibility" stands in stark contrast to Lavoisier's focus on measurable phenomena. For Dalton, empirical results about the solubility of gases were not intelligible unless accounted for in terms of the particles that constituted them.

Dalton's view of the particulate nature of gases also informed his ideas about consistent differences in solubility of different species of gas. He accounted for the variation in terms of differences in the physical characteristics of the gas particles: those gases that were composed of small, light particles would be more soluble than those consisting of larger, heavier particles.

Lavoisier had already discussed the particles of simple substances in his *Traité élémentaire* in 1789, and had ruled out the possibility of saying anything valid about them.

All that can be said upon the number and nature of elements is, in my opinion, confined to discussions entirely of a metaphysical nature. . . . I shall therefore only add upon this subject, that if, by the term *elements*, we mean to express those simple and indivisible atoms of which matter is composed, it is extremely probable we know nothing at all about them; but, if we apply the term *elements*, or *principles of bodies,* to express our idea of the last point which analysis is capable of reaching, we must admit, as elements, all the substances into which we are capable, by any means, to reduce bodies by decomposition.[21]

By branding speculation about the atoms of "elements" as metaphysical, Lavoisier effectively condemned it as meaningless; since the seventeenth century the label "metaphysical" had often been used to castigate natural-philosophical ideas for supposedly going beyond the evidence of direct experience. Dalton, of course, did not see things in that way.

In 1807, Dalton delivered a series of lectures in Edinburgh and Glasgow. The lectures presented what he now called his "doctrine of atoms." These atoms were to be understood as the "primary elements" of bodies. Unlike Lavoisier, Dalton thought that he could adduce empirical evidence that would support the transmutation of Lavoisier's operationally defined "elements or principles" into physically real atomic particles. Once again, gases were the route that he followed to arrive at this conclusion.

In the chemistry that had been established by Lavoisier, carbonic acid gas was called "carbonic oxide," a name based on the constituent simple substances, carbon and oxygen, that seemed to compose it. Given Dalton's conception of the physical nature of a gas, the only way that he could make sense of this chemical combination was to imagine that carbonic oxide, as a gas, must be made up of particles that were themselves compounds of these two basic substances. He already assumed that oxygen gas consisted of mutually repelling oxygen particles; to get carbonic oxide, therefore, he proposed that each particle of oxygen must be attached to a particle of carbon. The gas called carbonic oxide would then be understood as composed of mutually repelling *compound* particles, each consisting of one oxygen atom joined to one carbon atom. Dalton's choice of numbers—one oxygen joined to one carbon—was based on nothing more than a guess that the simplest possible ratios were the most likely (fig. 3.2).

This simple picture could quite easily be extended from cases of compound gases to cases of solid compounds, thanks to work that chemists in Continental Europe had recently produced to show what they claimed were the constant proportions of weight involved in the formation of chemical compounds. Lavoisier's classic example of the formation of water from hydrogen and oxygen, for example, had produced the experimental conclusion that eighty-five parts by weight of oxygen combined with fifteen parts by weight of hydrogen to yield one hundred parts of water. Other chemists had subsequently argued for constant weight ratios in the chemical formation of solid compounds made from solid reactants. That data, although

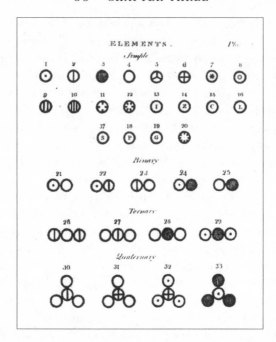

3.2. Dalton's atoms (the first four are hydrogen, "azote" or nitrogen, carbon, and oxygen) and their combination into the compound atoms (now called molecules) of chemical compounds; 21 is water, 22 is ammonia, etc. From Dalton's A New System of Chemical Philosophy (1808). Note that Dalton generally used the simplest number ratios for compounds: since only one oxide of hydrogen was known, water was assumed to be the simplest ratio of one oxygen atom combined with one of hydrogen, and not, as now, one-to-two. The more complicated ratios (as shown for ternary and quaternary) are the result of further combination that produces new chemical properties—thus, adding an oxygen to carbonic oxide (25) gives the "atom" of carbonic acid (28). Not until the 1860s, and the general acceptance by chemists of particular techniques for determining relative atomic weights, did the newer formula receive general acceptance.

by no means uncontested by prominent chemists at the time, was grist to Dalton's mill, and enabled him to extrapolate constant weight ratios to all cases of chemical combination. He could then say that these constant weight ratios were due to the combinations of constant small-number ratios of atoms, and that the actual measured weights were evidence of the relative weights of the atoms themselves. In one of the Edinburgh lectures, Dalton described his goal as being to reduce the system of chemistry "to a science of great simplicity, and intelligible to the meanest understanding."[22]

The classic statement of Dalton's ideas appeared in his book *A New System of Chemical Philosophy*, the first part of which was published in 1808. But Dalton's work by no means established chemical atomism as the foundation of chemistry. It was criticized on various grounds, including explicitly philosophical ones, for the rest of the nineteenth century, many chemists being more comfortable with laboratory measurements, and the generalizations based upon them, than with ideas of unobservable atoms. By turning Lavoisier's empirically defined chemical principles from digests of laboratory experience into physically real atomic elements, Dalton had rejected Lavoisier's chemical methodology. He appears to have done this because Lavoisier's approach, followed by many other chemists, did not serve Dalton's own attempts to make sense of a variety of physical and chemical phenomena. For Dalton, intelligibility rested on physical pictures of the world, not on a codification of laboratory practice.

CHAPTER FOUR

Design and Disorder:
The Origin of Species

I. Design in Nature

In late 1859, soon after the first publication of his *The Origin of Species*, Charles Darwin (1809–82) wrote to the eminent geologist Charles Lyell (1797–1875), "I have heard by round about channel that Herschel says my Book 'is the law of higgledy-piggledy.'"[1] John Herschel (1792–1871), the son of the astronomer William Herschel, was not only an astronomer himself, but also an influential philosopher of science whose opinion Darwin was bound to take seriously. Darwin was unsure as to what precisely Herschel had meant by his remark, but was in no doubt that it was not meant to be complimentary to the theory of natural selection.

Darwin's theory represents a remarkable new conception of what it meant to account for many features of organic nature in the nineteenth century. Rather than proposing novel answers to essentially the same, preexisting questions, Darwin attempted to translate those questions into new forms that radically altered their meanings. Darwin hoped that his newly reformulated questions would be regarded as illuminating the old ones from a new point of view, but the risk was that they would simply be rejected as missing the point—and if the latter, Darwin's new *answers* would become absurd or irrelevant. Which of those alternatives his contemporaries chose was fundamental to determining the responses to Darwin's theory. First among these reformulated questions was that of design.

The designfulness of organic nature was taken for granted in the world of British natural history in the 1830s, when Darwin first began to think about the problem of species and their interrelationships. In the eighteenth century, Linnaean natural history had assumed a God-given order behind the taxonomic relationships among plants and among animals. At the beginning of the nineteenth century, Cuvier's work had extended the philosophical legitimacy of zoological, and by implication also botanical, taxonomy by purporting to ground it in the requirements of functionality—what it took for an animal to be viable. The underlying assumption of Cuvier's approach to anatomy and classification, in other words, was that animals are functionally well organized. Such a view became especially popular in Britain in the first half of the nineteenth century because it chimed with a separate, well-established British intellectual tradition, dating from the later seventeenth century, known as natural theology.

Natural theology sought to identify features of the world that gave evidence of an intelligent and designful Creator. Stress on the immensely elaborate natural artifice of plants, animals, and even aspects of the heavens became a pronounced feature of much British natural philosophy during the opening decades of the eighteenth century, and remained strong well into the nineteenth; Isaac Newton had himself made arguments of this kind in his *Principia* and *Opticks*. Newton asserted that certain notable regularities in the organization of the solar system were of a kind that natural philosophy alone could not account for (his own version being the test, of course). For example, he noted that the planets all orbit the sun in roughly the same plane (the ecliptic), and that their orbits are all approximately circular. But there was no reason, given his mechanics and his gravitational law, why that should be the case: the planets might just as well orbit in a random variety of planes, following wildly elliptical paths differing greatly from circles. Newton reckoned, therefore, that the actual order of the solar system had not been produced by the action of blind natural laws, but must instead have been established deliberately by God, the source of intelligent design.

The leading natural theologian of the early nineteenth century was the Reverend William Paley (1743–1805), whose book *Natural Theology* first appeared in 1802 and went through countless editions thereafter. Paley focused primarily on organic creatures to show God's care in providing them with just the characteristics that they needed to survive: ducks conveniently possess webbed feet to aid their swimming; our own eyes, delicate as they are, are equipped with eyelids and eyebrows to protect them; marsupials have pouches to house their defenseless young, and so forth. The formal dimension of Paley's argument was simple, and powerful. He invited his readers to consider their own reasoning when coming across a watch lying on the ground. Surely, he urged, no one would regard this find as simply a randomly generated agglomeration of matter, like a pebble; instead, one would note on investigation that the object was of complex contrivance, and seemed to be wonderfully designed for the apparent purpose of running regularly and keeping time. The immediate inference such a discovery would prompt was that this object had been made by someone, an intelligent creator. Why, then, he asked, should one not draw the same inference from natural examples of elaborate designfulness? Paley summarizes his entire argument towards the end of his book: "Design must have had a designer. That designer must have been a person. That person is GOD."[2]

By the time that Darwin attended Cambridge University in the late 1820s, Paley's book was a required text for all its undergraduates. Darwin duly read it, and professed that it had given him "as much delight as did Euclid"[3]—evidently a mark of praise for the geometrical rigor of Paley's reasoning. The easy fit between this kind of natural theology and Cuvier's stress on adaptation of form to function (always, it was assumed, perfect) meant that, to British naturalists of the 1830s, it was practically an article of scientific faith that one of the outstanding features of organic beings was that they were remarkably well suited to their ways of life. This was one of the givens of experience that Darwin knew from the very outset would have to be taken into account in making sense of the diversity and interrelations of living beings.

II. Species and Design

After his return in 1836 from the lengthy voyage around the world of the HMS Beagle, Darwin began to focus on what he called "the species question." This was not a straightforward and obvious focus for his work: the question concerning species was one that had emerged from a complex of issues in taxonomy and paleontology, as well as other areas of natural history. Darwin's famous voyage had given him the opportunity to study the natural history of a wide variety of localities throughout the globe, especially in South America; but how he studied them was conditioned by the expectations that he brought to them. During his days as an undergraduate, Darwin had become enamored of such pursuits as the collecting of beetles, which was typical of the interests of contemporary English amateur naturalists. Inevitably, this hobby brought Darwin into contact with issues of taxonomy, since collecting natural specimens involved identifying them according to species. An eye for distinguishing a separate species from a mere variety was something that a budding naturalist like Darwin acquired as a matter of course. But that eye was not foolproof.

The dead finches that Darwin brought back to England from the Galápagos Islands (fig. 4.1) were, according to his own account, something of a revelation to him on this very score. He had acquired specimens from various of the islands, and while he noted slight differences between them, he had thought they were merely marks of different varieties, and that the finches were all of one species. When he returned to England, however, a respected avian taxonomist, John Gould (1804–81), told him that his specimens were in fact members of distinct species. But Gould's revelation to Darwin must not be taken at face value. What, after all, was a "species"? In the wake of Cuvier's work and its general acceptance, particularly among

4.1. Darwin's Galápagos finches, from his Journal of Researches *(1845): similar but distinct?*

British naturalists, species were regarded as stable natural kinds rather than flexible forms that could slowly transmute into new forms. Buffon's criterion, that two individuals counted as being of the same species if they were members of a common breeding community, played no part in the practice of taxonomy in Darwin's time: Gould could only look at dead specimens of the finches, and yet he had no reluctance in pronouncing them representatives of different species. On what basis could Darwin accept this claim? What did the concept of species mean to him at this time, as well as later, when he began to believe in transmutations that allowed the appearance of new species?

In fact, Darwin was more perplexed than convinced by Gould's assertion. It seemed to him that the distinction between varieties and species was perhaps less clear than trained taxonomists like Gould, whose job it was to draw such distinctions, took them to be. Darwin had the advantage at this time of not being an experienced taxonomist, and his own tendency towards sloppiness in classification began to turn, in his own mind, into a virtue. It was later in 1837, the year marked by Gould's taxonomic pronouncements, that Darwin became secretly persuaded that species transmuted into new forms, and that such a process fitted well the taxonomic claims of his colleagues. Species, he began to believe, were just highly differentiated varieties.

It is one of the remarkable facts of nineteenth-century natural history that the practices of taxonomists were not thrown into disruption by the eventual publication and acceptance of Darwin's ideas after 1859. Darwin was to rely on taxonomy for much of his argument in the *Origin of Species*, reinterpreting its meaning in the terms of branching trees of descent. He never paused to ask whether the very meaning of the category "species" might have been radically changed by his theory, in such a way that earlier taxonomic practices would have to be called into question. It would not have been in his interests to do so, because existing taxonomy provided him with valuable arguments. Taxonomists demonstrated lines of filiation between species; Darwin would simply explain what these connections meant in a new and, he hoped, more convincing way.

Darwin's endeavor involved two central themes. One, the most daring, was that new species emerged through transformation from older species—what was later called (although not in Darwin's *Origin*) "evolution." The other, of equal importance, was the principal mechanism that Darwin proposed to bring about transmutatory change in those groups of organisms that would produce the new species: natural selection. The doctrine, soon popularized by the British pop philosopher Herbert Spencer (1820–1903) as "survival of the fittest," held that individual organisms routinely display slight variations as compared with others of the same species—on a par with those by which we distinguish individual people, even siblings, from one another. Furthermore, these variations usually correspond to some greater or lesser ability of the organism to cope with its circumstances of life, whether being enabled by fractionally longer or stronger legs to flee predators more successfully, or being capable (as with some of Darwin's Galápagos finches) of reaching food secreted more deeply inside the bark of a tree owing to a longer beak. As a result of such small advantages, some individuals of any particular species would be more likely to survive in the "struggle for life" than their fellows. And since offspring tend to resemble their parents, the lucky survivors would probably pass along their advantages to their children. As a result, advantageous characteristics would proliferate in the population, and disadvantageous ones would tend to disappear. Hence John Herschel's "the law of higgledy-pigglety": nature shakes the weighted dice, and the weighting affects the likely outcome.

Darwin had introduced natural selection in the *Origin* as part of an overall argumentative strategy. He had begun with a discussion of domestic selection, the kind of selection that a human breeder of animals or plants performs when choosing which specimens to breed from. This focus enabled Darwin to show from everyday experience that individuals within a species really display a lot of variation, and how deliberate selection in breeding, with the intention of accentuating particular desired characteristics, could result in organisms often very different from their ancestors. One of his favorite examples was that of show pigeons, which were all generally admitted to have descended from the rock dove; and yet that common origin had not prevented the most diverse and extraordinary breeds from being produced over the course, in many cases, of historically recorded time. Darwin was able to cite similar changes in the course of no more than a couple of centuries in favored breeds of dog. Once he had established the possibility of such change, Darwin then suggested a means by which systematic selection of particular traits could occur naturally, without human choice and intervention. Natural selection was, then, meant to be a natural, and indeed materialistic, mechanism rooted in the everyday intelligibility of the work of stockbreeders and fanciers of show dogs. But the very fact that the term retained the word "selection" meant that the old natural-theological sense of intelligent designfulness still lurked in the background. Apparent designfulness was an important part of what Darwin wanted to explain, after all, and as long as he took its existence for granted his metaphor of selection was bound to be understood by others as implying a selective agent.

III. Time and Imagination

A crucial feature of natural selection was that it would be capable of creating a multiplicity of new species from common ancestors only if it had an unimaginably long period of time in which to operate. In order to produce such diverse species as squids and elephants while relying only upon tiny individual variations and a very insecure sorting mechanism, natural selection required an immensely old earth. Because Darwin had begun his scientific career as a geologist he knew well the issues involved, but he could not prove that the earth really was as old as he wanted it to be—indeed, he could not even calculate such a vast age.

He tried, in one notable instance, to produce a specific number for the age of a particular geological process, simply to show how large such numbers could be. In the *Origin*, Darwin draws attention to a topographical feature found in southeastern England: the lowland area between the parallel hill regions of the North Downs and the South Downs known as the Weald. Darwin estimates the rate at which these chalk formations might have been eroded by the action of the sea so as to create the twenty-two-mile gap between the two Downs. Making what he represents as a conservative estimate of the rate at which a high chalk cliff face might be worn away by the sea (one inch per century), Darwin calculates that the entire period required for the formation, or "denudation," of the Weald would have been 306,662,400 years. He then adds considerations that allow him to say that "in all probability a far longer period than 300 million years has elapsed."[4]

> During each of these years, over the whole world, the land and the water has been peopled by hosts of living forms. What an infinite number

of generations, which the mind cannot grasp, must have succeeded each other in the long roll of years![5]

Even here, in other words, having produced an apparently precise number for a geological period, Darwin hurried to deny its accuracy. Instead, the number simply formed the pretext for conjecturing a "much longer period," quickly followed by talk of "an infinite number" of organic breeding generations. Darwin's general dislike of precise numbers was in this case warranted: others soon questioned his figures for the denudation of the Weald. A reviewer of the *Origin* wrote that "in this delicate part of his case he has committed some manifest errors."[6] Darwin soon confessed his lack of circumspection in print, remarking: "I own that I have been rash and unguarded in the calculation."[7]

All Darwin really cared about was that the earth should be older than any limit his opponents might wish to place on it. After almost two decades of debate on the subject, Darwin's son, the scientist George Darwin (1845–1912), wrote the following to his eminent fellow physicist William Thomson (1824–1907; later Lord Kelvin), who advocated a relatively short age for the earth based on physical estimates of its rate of cooling:

> I have no doubt however that if my father had had to write down the period he assigned at that time [i.e., when he wrote the *Origin*], he [would] have written a 1 at the beginning of the line & filled the rest up with o's. Now I believe that he cannot quite bring himself down to the period assigned by you, but does not pretend to say how long may be required.[8]

This candid description of Charles's view of geological time in relation to his theory captures what is perhaps its central characteristic: Darwin viewed prodigious quantities of time as an unlimited frame within which to allow his imagination to create the actual organic world from the simplest beginnings. The theory of natural selection, despite all the circumstantial (and fully admissible) evidence that Darwin brought forward in its support, was fundamentally a thought-experiment. Again and again in the *Origin,* Darwin invites his reader to imagine how natural selection might have brought about the organs or behaviors that are observed in nature—the mere *possibility* that it might have done so is what he needs in order to counteract the assumptions of natural theology, such as William Paley's argument that the eye gave unquestionable evidence of an intelligent Creator. Darwin describes in the *Origin* how the eye might have developed over countless generations, as layers of transparent tissue gradually came to form an ever more perfect, and selectable, optical instrument.

> We must suppose each new state of the instrument to be multiplied by the million; and each to be preserved till a better be produced, and then the old ones to be destroyed. In living bodies, variation will cause the slight alterations, generation will multiply them almost infinitely, and natural selection will pick out with unerring skill each improvement. Let this process go on for millions on millions of years; and during each year on millions of individuals of many kinds; and may we not believe that a living optical instrument might thus be formed as superior to one of glass, as the works of the Creator are to those of man?[9]

Ironically, this passage occurs just after remarks by Darwin on how anyone understanding the *Origin* correctly would have to "admit that a structure even as perfect as the eye of an eagle might be formed by natural selection. *His reason ought to conquer his imagination.*"[10]

Darwin's attempts to persuade, however, use reason to explore ideas that have first been conjured up by the imagination. He uses indefinitely large numbers—numbers that are allowed to be as large as you like—to do just that. Vast lengths of time, and vast numbers of individual organisms—so vast, indeed, as to beggar the very imagination that Darwin tries to downplay—lend a kind of sublimity to his vision of the history of life; he uses imagination in such a way as to transcend it. The role of reason is really to enable a leap from ordinary imagination to the literally unimaginable.

The sole illustration in the *Origin of Species* illustrates the same

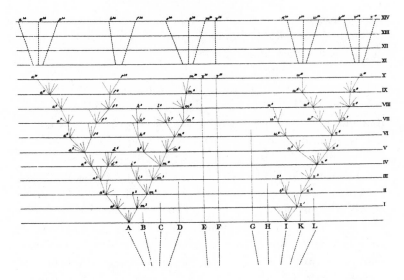

4.2. Diagram of descent with modification, from the Origin of Species *(1859). Time progresses* up *the page.*

point (fig. 4.2). The horizontal lines in this diagram represent particular points in the earth's history, going from oldest (at the bottom) to most recent (at the top); the branching lines that trace up the diagram stand for the ever-multiplying and diverging forms of life, whereby new varieties, species, and eventually even broader categories such as genera and families come into being through the action of natural selection. Darwin comments: "The intervals between the horizontal lines in the diagram, may represent each a thousand generations; but it would have been better if each had represented ten thousand generations."[11] The precise intervals of time, that is, do not really matter; they must simply be immense.

Why should anyone have been convinced by numbers, especially numbers that were deliberately imprecise? What kind of understanding could they have generated? What was intelligible about the unthinkably large? Darwin frequently used the power of the imagination to allow his readers to envision the possibility of implausible things; here, however, he wanted unrestricted time to show a way in which the incapacity of the imagination to grasp something should not count as a mark of its impossibility.

Darwin's appeal to the imagination of his readers was sometimes quite extraordinary, and appeared so to his contemporaries. The *Origin* contains such passages as this: "I can, indeed, hardly doubt that all vertebrate animals having true lungs have descended by ordinary generation from an ancient prototype, of which we know nothing, furnished with a floating apparatus or swimbladder."[12] This statement appears after some discussion of the comparative anatomy of vertebrates, and draws on the work of conventional, non-evolutionist physiologists: "All physiologists admit that the swimbladder is homologous, or 'ideally similar,' in position and structure with the lungs of the higher vertebrate animals: hence, there seems to me to be no great difficulty in believing that natural selection has actually converted a swimbladder into a lung, or organ used exclusively for respiration."[13] In effect, Darwin asks his readers to conceive, as a reality, that their own lungs were once swim bladders inside the fish that were their direct ancestors. A more intimate kind of imagining could scarcely have been proposed; appropriately, Darwin frames it in his typical terms of "there seems to me," and "I can hardly doubt."

Darwin's great critic, the eminent anatomist Richard Owen (1804–92), noticed and disapproved of this style: "we do not want to know what Darwin believes & is convinced of, but what he can prove."[14] Owen would not allow Darwin to proceed in the only way he actually could: to present his arguments in such a way as to show, first, the plausibility of the theory, and only then to develop links to factual "proof" as evidence of its truth. Darwin wanted his readers to follow him in seeing the world, and, by implication, themselves, in a new and unorthodox way. His theory attempted to *explain* the facts rather than to be supported by them; consequently, he had to present it ready-made—to show what a world in accordance with his theory would be like. Time, for Darwin, was like a railway line, always extendable: when the line sought to reach a given station, he always wanted there to be more track available to make it possible. If the remote past, in which lungs were swim bladders, could be placed so far off that it could not be seen directly, the apparent im- plausibility of the connection to the present (from swim bladders to his readers' own lungs) could at least be held in abeyance, giving Darwin a breathing space of credulity in which to marshal evidence for his overall account of things. He wanted his readers to indulge him with a willing suspension of disbelief until they were ready pos- itively to believe him.

Darwin seldom invoked imagination itself as a help in his argu- ments; he preferred to say such things as "I can see no difficulty" in the processes of change that he postulated. One especially striking example led him into some trouble. It concerned the possibility of a bear transmuting into a whale, and it required a particularly willing imagination on the part of his reader.

> In North America the black bear was seen by [the eighteenth-century travel writer Samuel] Hearne swimming for hours with widely open mouth, thus catching, like a whale, insects in the water. Even in so ex- treme a case as this, if the supply of insects were constant, and if better adapted competitors did not already exist in the country, I can see no difficulty in a race of bears being rendered, by natural selection, more and more aquatic in their structure and habits, with larger and larger mouths, till a creature was produced as monstrous as a whale.[15]

When Richard Owen wrote a review of the *Origin* in *The Edinburgh Review*, he singled out this passage for especial ridicule. He compared it to the generally disreputable arguments of the early nineteenth- century French naturalist Jean-Baptiste de Lamarck (1744–1829), a renowned advocate of transformist views on the origin of species. "We look, however, in vain for any instance of hypothetical transmu- tation in Lamarck so gross as the one above cited."[16] The Cambridge geologist Adam Sedgwick (1785–1873), firmly set against Darwin's ideas and devoted to natural theology's "argument from design," was even more dismissive in his review in *The Spectator:*

In some rare instances, Darwin shows a wonderful credulity. He seems to believe that a white bear, by being confined to the slops floating in the Polar basin, might in time be turned into a whale; that a lemur might easily be turned into a bat; that a three-toed tapir might be the great grandfather of a horse; or that the progeny of a horse may (in America) have gone back into the tapir.[17]

In the second edition of his book, published only six weeks after the appearance of the first, Darwin took the opportunity to alter passages that had rapidly been brought to his attention as inaccurate or else dangerously overstated.[18] The passage on the bear now appeared in a much truncated form, reading in its entirety: "In North America the black bear was seen by Hearne swimming for hours with widely open mouth, thus catching, almost like a whale, insects in the water."[19] The inference was now left to the reader. The imagination was evidently being stretched too far for some tastes: a correspondent of Darwin's said that the example "simply made me laugh."[20]

The point of Darwin's insistence on using indefinitely large numbers, whether for generations, individuals, or periods of time, was precisely to combat sneering remarks like Sedgwick's, "that a three-toed tapir might be the great grandfather of a horse." In the same way, the bishop of Oxford, Samuel Wilberforce (1805–73), supposedly asked Thomas Henry Huxley (1825–95), Darwin's great champion in the controversy over the *Origin*, whether he preferred to be descended from an ape on his grandfather's or his grandmother's side.[21] The ridicule rested on the manifest absurdity of transmutation in the face of common knowledge that species routinely reproduce their own kind. Darwin's uncountable numbers were a way of trying to avoid such an impression by placing unfathomable distance between ancestors and their transmuted descendants.

Strategy in writing is one thing; success another. Large numbers and the attempt to lull readers into identifying with Darwin's own convictions frequently failed. Some critics denied him his numberless numbers; others, his sense of the believable. The physicist William Thomson reduced the age of the earth to more manageable proportions by putting a value on it; even with a wide margin of uncertainty, the maximum figure that Thomson allowed was four hundred million years. This was too little for Darwin's needs, because his vision of natural selection involved an unquantifiable slowness in the production of its effects.

Darwin's use of time resembled the stellar astronomer's use of space, which could similarly be extended as far as was needed. The historian and former paleontologist Martin Rudwick has referred to nineteenth-century views of the earth's geological and organic past, inconceivably far back before humanity ever appeared on the scene, as "deep time," by analogy with "deep space," and the term is apt. Something so vast, "which the mind cannot grasp," was crucial to Darwin's mechanism for the transmutation of species, since the process itself, unlike its results, could never be caught in action. The initial plausibility of Darwin's theory relied on the mind's inability to relate together the time required for natural selection to produce its effects and the time that human beings actually experience. Intelligibility was to be produced by holding at bay unimaginability. A hostile reviewer of the *Origin* illustrates the importance of Darwin's strategy quite clearly. Thomas Vernon Wollaston (1822–78) asked, in regard to the implication (dodged by Darwin in the *Origin*) of the comparatively recent evolution of the human form,

> how is it that *no traditions whatsoever* bearing on the previous and more simple conditions of the human structure (immediately before it attained its climax of perfection) have ever been extant; for it is quite inconceivable that so radical an organic change could have been slowly brought about without, at the least, *some vague tradition of it* having become *a fact of the human mind.*[22]

Darwin had tried to distance inconceivability by the interposition of time beyond measure.

The issue of time was not so pressing for some of Darwin's leading supporters. Huxley, the chief among them, was an evolutionist, but not a strong natural selectionist. Huxley had already caused something of a scandal in 1856 by telling his fellow anatomists that they were not really doing what they thought they were. Cuvier's approach had long been a watchword in British anatomical practice: the anatomist was not only to describe an anatomical feature, but also, at the same time, to account for its structure in terms of that feature's function. Huxley said that, in reality, anatomists were really only concerned with morphology, that is, with description of structure, and that their functional explanations for that structure were simply tacked on as an afterthought, playing no real role in the anatomist's account. Huxley thought that animal morphology was not in fact very closely aligned to function, being himself unconvinced by both Cuvier's arguments and natural theology's argument from design. Animals simply made the best use they could of what features they happened to possess. So for Huxley, natural selection, which Darwin had proposed as a way of giving a natural explanation for the apparent designfulness of organic beings, was relatively unimportant, useful only for weeding out grossly unfit variants. Huxley cared chiefly about transmutation, or evolution, together with the evidence for it in the fossil record (fig. 4.3). The age of the earth was not, therefore, a crucial factor for him: if the earth was said to be younger than natural selection required, then evolution, driven by some other mechanism, had simply proceeded faster than Darwin supposed.

Huxley supported Darwin nonetheless. Darwin's critics were less flexible. They objected to the doctrine of evolution itself, and their attacks on natural selection served as a means to that end. Darwin faced many problems in marshaling general acceptance for the various parts of his elaborate theory.

IV. Variation and Selection

The Origin of Species gave Darwin enormous trouble to write. It appeared in 1859 as the outcome of more than two decades of research, thought, and writing, its eventual publication being forced on

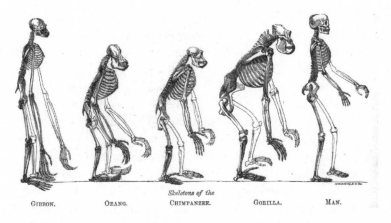

GIBBON.   ORANG.   Skeletons of the CHIMPANZEE.   GORILLA.   MAN.

4.3. Man and apes, from Huxley's Man's Place in Nature *of 1863. Huxley says that he has shown the gibbon "twice as large as nature."*

Darwin by the threat of being forestalled by the much younger Welsh naturalist Alfred Russel Wallace (1823–1913); without that threat, a full account would have taken even longer to appear in print.

Historians have often remarked that Darwin's book appeared at a time and in a place well suited for it. Writing about a typical Darwinian remark concerning the "war of nature," Charles Gillispie has observed, in his *Edge of Objectivity:* "None but a Victorian Englishman could have written those words."[23] The picture that Darwin put forward, of a nature in which individual organisms are constantly struggling to survive, corresponded well to a prominent strain of thought in Britain in the first half of the nineteenth century that owed a lot to the Reverend Thomas Malthus (1766–1834).

Malthus's *Essay on the Principle of Population* (1798, with many subsequent revised editions) put forward the view that overscrupulous poor relief could only make poverty worse—a popular argument in some quarters in Britain in Darwin's time, and related to the notorious Victorian workhouses that Charles Dickens wrote about. Malthus's central idea concerned the tendency of human populations to increase at too rapid a rate, and the consequent necessity of ensuring restraint in population growth. If population increases too rapidly, it will soon exceed the ability of the country to feed it. Families should therefore be persuaded not to grow too large. But poor families, Malthus said, tend to produce lots of children in spite of the poverty that ought to dissuade them: this unfortunate state of affairs was directly encouraged by the routine provision of charitable poor relief, which undermined a major incentive to limit family size. Essentially, Malthus feared a population explosion that would create widespread poverty and hunger when the growing population outstripped available resources; the only way to avoid this catastrophe was to institute reproductive restraint (Malthus's favorite technique was something he called "moral restraint"), and the limiting of poor relief was a direct means of trying to achieve it by force. Malthus's book was a contribution to what was known as "political economy," and this intellectual pretension ensured a wide readership among the well-to-do, who often welcomed its political message.

Darwin read Malthus in September 1837 while working feverishly on his speculations about species. He had convinced himself by that time that transmutation of species really did occur, but still sought what he called a "natural" mechanism to explain it. Malthus had emphasized the tendency of a population to grow exponentially, since each (human) breeding pair typically produces more than two children (just two would suffice to keep the overall population constant since they will compensate, eventually, for the death of their parents). The inevitable result of this typical overbreeding would therefore be an ever-expanding population and eventual shortage of resources, once the population had reached its maximum supportable size. Darwin took from Malthus's argument the central idea of competition among individuals of the same species. Malthus had spoken only of the human species, but the same phenomenon ought, Darwin thought, to apply to any species. Thence arose in his mind the struggle for survival, whereby individuals of the same species would be in competition with each other for food, mates, shelter, and all other necessities for their successful survival and reproduction.

Malthus showed where the pressure for continual competition came from: the tendency of the population constantly to push the limits of its sustainable size, such that not all those individuals born could possibly survive to reproductive age. That was Darwin's engine to power the operation of natural selection.

The parallel with Malthusian political economy (which Darwin openly acknowledged) was something that was taken up by others, not always sympathetic to Darwin's ideas. In 1860, the Irishman Samuel Haughton (1821–97), of Trinity College, Dublin (then a bastion of the Protestant British establishment), compared Darwin's theory to that of Lamarck, whose views Haughton held to be typical of the French in "the vivacity and perception of the ridiculous belonging to his nation."[24] Darwin, an Englishman, by contrast "firmly believes his theory [of natural selection], and, with a confident faith in the power of food and comfort, equally characteristic of his country, elevates the desire to supply the stomach into a law of sufficient force to convert an eel into an elephant, or an oyster into an orangoutan."[25] But not only opponents of Darwin pointed out the parallels with British theories of political economy. William Benjamin Carpenter (1813–85) was a leading physiologist, and in his review of the *Origin* explained to his readers Darwin's view of the consequences of a change in environmental conditions for some particular variety, or "race," of a species. In such circumstances, "the race must either be capable of adapting itself to that change, or it must succumb.

> In the one case, the original form will give place to some modification directly proceeding from it by genetic descent; in the other, it will be superseded by some rival form derived from a different ancestry, which presses in and occupies its place; just as we see, in the social battles of life, that the families of our older aristocracy hold their ground, or are displaced by the *parvenus* whom they regard as their natural enemies, in proportion as they either adapt themselves to the spirit of the age and take advantage of its requirements, or as they hold tenaciously to their time-

honoured customs, and refuse to profit by any thing that shall lower them in their own artificial scale of dignity.[26]

Such comments by contemporaries show a fair degree of cultural recognition for Darwin's central concept of natural selection. But the different attitudes towards it of opponents and supporters suggest that the intelligibility of natural selection rested to a large degree on sympathy for the political and social picture that this recognition called to mind. Haughton disliked what he represented as a rather lowering parallel between the mundane matters of political economy and the ineffable splendor of the natural world's variety. Carpenter, by contrast, seems to have welcomed the simplicity of an explanation the clarity of which could be brought home by such well-known and easily understood social phenomena. Sympathizers with conservative social thought generally disliked evolution as a doctrine, and therefore sought grounds to reject the plausibility of natural selection; those who liked the idea of social mobility and the doctrines of nineteenth-century liberalism, with its fondness for laissez-faire economics, tended to like evolution as well as natural selection, which presented an image of a competitive economy in nature itself. Political views tend to be visceral, and responses to Darwin exhibited a similarly visceral feeling regarding his theory's intelligibility—as to whether it made perfect sense or was simply ridiculous.

One of Darwin's more important champions well understood the importance to him of natural selection. The botanist Joseph Hooker (1817–1911), now at Kew Gardens, had been the first reader of Darwin's earliest developed version of his theory in 1844, and by the time of the *Origin*'s appearance he was ready to commit to it. Hooker's review of the book in late 1859, in the unlikely venue of the *Gardener's Chronicle,* displayed a full understanding of Darwin's project. Darwin, he wrote,

has endeavoured to invent and prove such an intelligible rationale of the operation of variation, as will account for many species having been developed from a few in strict adaptation to existing conditions, and to show good cause how these apparently fleeting changelings may by the operation of natural laws be so far fixed as to reconcile both the naturalist and the common observer to the idea, that what in all his experience are immutable forms of life may have once worn another guise.[27]

However, a German reviewer, the zoologist and paleontologist Heinrich Georg Bronn (1800–1862), while finding a degree of intelligibility in Darwin's arguments, refused to draw Darwin's own conclusions: "Though it [i.e., the theory] is conceivable, it still remains undecided whether all organisms from the simplest filament to those that are intricately constructed like the butterfly, snake, horse, etc. can be the production of just blind natural forces!"[28] Conceivability was all very well, but it was no evidence of historical truth.

Others called into doubt the theory's conceivability itself, both through their explicit criticisms and their evident misunderstandings. Another British reviewer, writing in *The Geologist*, allowed that

> [w]e should be more inclined to refer the modifications which species of animals or plants have undergone to the direct will of God, for it seems difficult to conceive how a being totally ignorant of its own structure or conditions of living should so commence modifying its structure, form, or habits, as to adapt not itself, but successively its progeny to new forms and conditions of life.[29]

The irony of this observation lies in the fact that Darwin's use of natural selection was a deliberate attempt to develop a natural explanation of adaptation, of apparent "designfulness," in place of its usual attribution to God's direct creative agency.

The very term "natural selection" gave Darwin some trouble, to the extent that he soon expressed the wish that he had used some other phrase that avoided the impression of conscious intent given by the word "selection." His correspondent William Henry Harvey (1811–66), in Dublin, was the person who drove him to it. "[I]t seems to me," Darwin complained to Harvey, "that you do not understand what I mean by Natural Selection." Darwin went on to note: "The term 'Selection' I see deceives many persons; though I see no more reason why it should than *elective* affinity, as used by the old chemists."[30] He repeated the wish to Lyell, naming Harvey as the problem, and followed through with a short discussion of the matter in the *Origin's* third edition. "Several writers have misapprehended or objected to the term Natural Selection," he began.

> In the literal sense of the word, no doubt natural selection is a false term; but who ever objected to chemists speaking of the elective affinities of the various elements?—and yet an acid cannot strictly be said to elect the base with which it in preference combines. It has been said that I speak of natural selection as an active power or deity; but who objects to an author speaking of the attraction of gravity as ruling the movements of the planets? Everyone knows what is meant and is implied by such metaphorical expressions; and they are almost necessary for brevity.[31]

The geologist Adam Sedgwick disliked Darwin's theory on a number of grounds, but a central theme to which he kept returning was that of morality. For Sedgwick, organic nature was a prime source of evidence for God's design, and natural history a guide to right and wrong. "There is a moral or metaphysical part of nature as well as a physical," he wrote to Darwin after having read the *Origin*. "'Tis the crown & glory of organic science that it *does* thro' *final cause*, link material to moral; & yet *does not* allow us to mingle them in our first conception of laws, & our classification of such laws whether we consider one side of nature or the other."[32] Sedgwick used the traditional label "final cause" as a standard term to refer to teleology, or purposefulness. The usual pre-Darwinian way of explaining features of the organic world, of course, had stressed purposes and goals so as to invoke God's intentions in creating things as He did; Sedgwick clearly saw this as a defining characteristic of natural history, not a dispensable one. "Passages in your book . . . greatly shocked my moral taste," wrote Sedgwick.

"Poor dear old Sedgwick" (as Darwin referred to his former Cambridge mentor in private letters) would not accept evolution or natural selection because they violated his most firmly held Christian conceptions of what the study of organic nature was really about. Near the end of the *Origin*, Darwin looked "with confidence to the future, to young and rising naturalists, who will be able to view both sides of the question with impartiality."[33] And yet Darwin himself could not entirely abandon the moral and religious overtones that were so important to Sedgwick. The *Origin* closed with the following famous lines:

> There is grandeur in this view of life, with its several powers, having been originally breathed into a few forms or into one; and that, whilst this planet has gone cycling on according to the fixed law of gravity, from so simple a beginning endless forms most beautiful and most wonderful have been, and are being, evolved.[34]

By 1871, the dust had largely settled on Darwin's claims about the reality of evolution, and his book *The Descent of Man* appeared with no great sensation. Everyone in the British scientific community and beyond had long assumed that his arguments for evolution were intended to apply to human beings. But accepting natural selection as its mechanism was another, and unresolved, matter. Throughout the 1860s a large number of objections to it had been formulated besides William Thomson's physical calculations of the earth's age, which seemed to leave insufficient time for natural selection to do its job. One of the most difficult challenges concerned the slight, everyday variations that Darwin proposed as the raw material of natural selection: many naturalists could not see such marginal differences being preserved, even when favorable, over several generations. The prevailing view of biological inheritance at the time, shared by Darwin himself, saw it as proceeding through the blending of the characteristics of the parents. So the first generation after the appearance of a favorable random variation in one individual would be diluted in that individual's offspring, since there was no reason to suppose that its mate would happen to share that same new, randomly appearing trait. Subsequent generations would see it diluted still further, until it had faded away altogether. This problem required Darwin to invent ever more contrived means whereby the dilution could be avoided, with the net result that few of his scientific colleagues became convinced that natural selection acting on slight variations was in fact the primary means by which evolution had occurred. But evolution itself, as a supposed descriptive fact, was much more easily accepted.

The theory, or assumption, of so-called "blending inheritance" posed a serious challenge to Darwin's natural selection, and was one of the sharpest arguments against accepting his evolutionary ideas. But there is an important sense in which any such argument merely expressed, in a formal way, a basic conviction of natural selection's implausibility. The intelligibility of a purely naturalistic explanation for the diversity and apparent adaptedness of organic beings remained a vexed question for far longer than Darwin would have wished, because teleological design remained for many people a central perception of the organic world. After 1900, development of the gene theory, based on the idea of "unit characters" first proposed by Gregor Mendel (1822–82), opened a way for natural selection to sidestep objections based on blending inheritance. Nonetheless, throughout the twentieth century the "law of higgledy-pigglety" remained one that a number of life scientists continued to find grounds for questioning: Stephen Jay Gould challenged the presumption of high degrees of adaptation as the central feature of evolution in his theory of "punctuated equilibria," in which evolution occurs in brief bursts separated by long periods of relative stasis; other people, including a small number of scientists, continued to find religious as well as nonreligious grounds for maintaining design as a central feature of the organic world.

CHAPTER FIVE

Dynamical Explanation:
The Aether and Victorian Machines

Nineteenth-century Britain was dominated by the realities of the industrial revolution and the commercial organization of society. The factory mode of production, with its attendant social implications, treated workers increasingly as interchangeable cogs in a machine, while machines themselves became ever more important in industrial enterprises. Most important among those machines was the steam engine, a source of power for industrial machinery that had been perfected by James Watt in the late eighteenth century. It provided a kind of model for British physical worldviews in the nineteenth century—a productive mechanism that linked the forces of nature to the mass production of the factory.

The great names among the Victorian physicists—James Prescott Joule (1818–89), William Thomson, W. J. Macquorn Rankine (1820–72), James Clerk Maxwell (1831–79), Peter Guthrie Tait (1831–1901), as well as many others—tended to be mathematically trained (Joule less so than most). But above all, they had in common an approach to physical problems that took the perspective of the mechanical engineer. Many were trained at Cambridge University, which was renowned for its state-of-the-art mathematics, including a particular concern for quantitative physical problems. But their mathematical expertise was not their most remarkable feature—there had been plenty of French mathematical stars, and there were many Germans too. Instead, the British physicists were characterized by their dislike of a particular physical concept, one that lent itself beautifully to mathematical management and that was perfectly accepted on the Continent: action at a distance. The treatment of gravitational attraction (and also electrical attraction) as though it were a matter of bodies exerting force on one another with no material intermediaries to transmit it had long been a standard part of the mathematician's toolkit. Yet the physicists of the great Victorian industrial world disliked it viscerally. They wanted their physics to be about things, not about abstract mathematical symbols, and the kinds of things they liked were the sort that could be hit with a hammer or poured into a drum. Those things made sense.

Near the beginning of the twentieth century, the French physicist and historian of science Pierre Duhem made some pointed observations on the subject. He wrote, with reference to a work by the British physicist Oliver Lodge (1851–1940):

> Here is a book intended to expound the modern theories of electricity and to expound a new theory. In it there are nothing but strings which move around pulleys, which roll around drums, which go through pearl beads, which carry weights; and tubes which pump water while others swell and contract; toothed wheels which are geared to one another and engage hooks. We thought we were entering the tranquil and neatly ordered abode of reason, but we find ourselves in a factory.[1]

Duhem explained this British peculiarity as the result of imagination triumphing over reason (the French, he thought, preferred reason). But for the British, it worked as a vehicle for intelligibility.

I. Lines of Force and Mechanical Explanations

The great experimentalist Michael Faraday (1791–1867) performed his groundbreaking work on electricity and magnetism, which included the invention of the electric motor and the dynamo, from the 1820s through to the 1850s. In that period, he developed the idea of

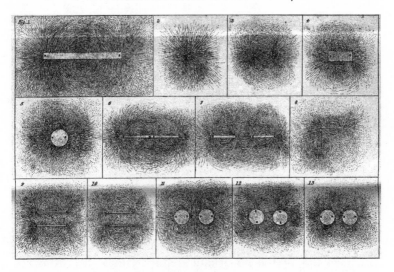

5.1. Faraday's own depiction of the arrangement of iron filings in the vicinity of magnetic poles. By this time, he had come to regard their dispositions as disclosing the existence of independently existing magnetic "lines of force," to which the filings orient themselves. From Philosophical Transactions, *1852.*

"lines of force" as a way to make sense of magnetic and electrical forces in the space around magnets and current-carrying wires. In the 1830s, Faraday had used this idea to conceptualize the three-dimensional reality of electrical and magnetic behaviors, but at that time he made no claim that the "lines" were literally present in space: they were simply a helpful way of thinking about experimental setups. In the 1840s and '50s, however, Faraday became convinced that these lines were real things, not just convenient fictions.

Faraday came to explain electrical lines of force in terms of chains of polarized particles in the material medium between a positively and a negatively charged body. Magnetic lines of force gave him more difficulty, since there was no evidence that the particles of matter (atoms, or molecules) could be regarded as magnets capable of forming chains comparable to the electrical ones (fig. 5.1). Instead, Faraday decided that these mysterious, but physically real, lines of magnetic force were due to something that he described as a "condition of space."[2] In other words, there was something qualitatively special about a region of space that contained magnetic force lines, although what that specialness consisted of he could not say in any noncircular way. It was enough for him that he could imagine the lines' existence around his experimental apparatus, and that he seemed to be able to detect and manipulate them. Faraday's lines of force possessed a reality, an intelligible existence, that was hands-on and tangible.

It was of less concern to Faraday that he could not specify in more theoretical terms what the lines in themselves were. But one of their clearly theoretical aspects was that, because they were a means of transmitting force through space, that transmission should be expected (thought Faraday) to take time. Like one of the most important physicists of the nineteenth century, James Clerk Maxwell, Faraday valued remarks that had been made by Isaac Newton in a letter written at the end of the seventeenth century but only published in the 1830s. The relevant passage from the letter was discussed in chapter 1, section II, above; it concerns the unintelligibility of action at a distance. In a paper of 1855, Faraday commented on the "strong conviction expressed by Sir Isaac Newton, that even gravity cannot be carried on to produce a distant effect except by some interposed agent fulfilling the conditions of a physical line of force."[3] Newton's authority regarding the impossibility—indeed, the inconceivability—of true action at a distance was useful to Faraday as a further means of establishing the physical reality of his lines of force. It also suited his assumption that gravitational force should be understood in the same general way as those of electricity and magnetism.

Many of Faraday's fellow British physicists fully sympathized with his rejection of a literal action at a distance: for Maxwell and others, it was not an acceptable way of explaining the exertion of force across space. So while Newton's remarks on the subject provided welcome support, Faraday's colleagues needed little convincing; they, like Faraday himself, were unable to conceive of any way in which a force might be exerted between distant bodies unless there was a means for transmitting that force across the gap. Faraday's ideas on force lines were unusual and controversial; more particularly, his concrete experimental sense of physical field lines was not shared by his more theoretically and mathematically conventional peers. They wanted to know, if the field lines truly existed, what they *were*. And the terms in which they wanted to know it were mechanical.

William Thomson, the physicist who was to give Darwin such a hard time over the age of the earth, strongly adhered to the view that making sense of electromagnetic phenomena should involve a mechanical model. The model would represent the material transmission and exertion of forces between bodies; this model should also, in principle, be mathematizable. Thomson explained his perspective in a lecture at Glasgow in 1846, his first as Professor of Natural Philosophy at the university:

> Every phenomenon in nature is a manifestation of *force*. There is no phenomenon in nature which takes place independently of force, or which cannot be influenced in some way by its action; hence mechanics has application in all the natural sciences; and before any considerable progress can be made in a philosophical study of nature a thorough knowledge of mechanical principles is absolutely necessary. It is on this account that mechanics is placed by universal consent at the head of the physical sciences. It deserves this position, no less for its completeness as a science, than for its general importance.[4]

Thomson continued by declaring that from "a few simple, almost axiomatic principles, founded on our common experience of the effects of force, the general laws which regulate all the phenomena, presented in any conceivable mechanical action, are established." As Thomson's biographers, Crosbie Smith and Norton Wise, put it, for Thomson "[n]atural philosophy . . . in effect meant *mechanical philosophy*."[5]

This conception of physical explanation was already widespread among British physicists and mathematicians in the 1830s and '40s.

One of its clearest expressions appeared in work by mathematicians who were trying to create a coherent view of the behavior of light. During the first two decades of the nineteenth century, the Frenchman Augustin Jean Fresnel (1788–1827) had produced a wave theory of light that soon received general acceptance. Instead of light being regarded as the effect of particles flying through space, which was Newton's view, Fresnel understood it in the terms of transverse (up-and-down) waves. Fresnel's theory explained numerous optical phenomena, such as the diffraction of light, that were harder to account for on a particle theory (fig. 5.2).

The acceptance of this wave theory as an adequate description of light propagation, however, created among British optical theorists an interest in providing a physical explanation for it. Waves, after all, needed a medium to carry them: sound waves could not exist in a vacuum, and waves on the surface of a pond could not exist without the pond's water; similarly, light waves needed a medium to be waves *of*. However, a special difficulty resulted from the theory's requirement that light waves be transverse waves, like those on a pond, even though they had to travel *through* the medium rather than over its surface. The waves of sound were longitudinal pressure waves through the fluid air, but light waves, with peaks and troughs that oscillated back and forth, placed difficult requirements on the kind of medium that could carry them. Any medium that sustained transverse waves would have to be, in effect, *wobbly*—an elastic solid like a block of gelatin. If it were a fluid medium like the air, any motion introduced into it would spread out through the medium in all (three-dimensional) directions rather than wiggling back and forth.

Mathematicians such as George Green (1793–1841), James MacCullagh (1809–47), and George Stokes (1819–1903) developed theoretical accounts of a medium suitable for carrying these transverse light waves. Their work possessed varying degrees of explanatory, as opposed to just mathematical descriptive, content. But Stokes in particular aimed at developing a theory of this medium, called the luminiferous (light-bearing) aether, that would give it real, self-consistent mechanical properties capable of accounting physically

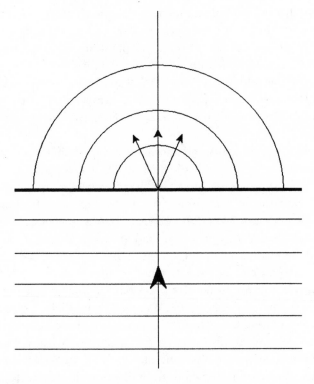

5.2. Diffraction of light refers to the tendency of light rays to spread out slightly from their straight-line path when they pass close by opaque objects; it is why shadows do not have entirely sharp edges even when produced by a point light source. Fresnel's wave theory of light explained this phenomenon in terms of the tendency of waves to spread outwards.

not just for the behavior of light, but also for electrical and magnetic forces. Thomson was much taken with Stokes's work, and adopted a similar approach. Of course, the issue of how ordinary bodies managed to pass through this postulated solid aether without resistance remained something of a puzzle. But the central intention was clear: natural phenomena would be understood as the outcome of material, mechanical interactions. Those were the kinds of processes that Thomson regarded as having inherent intelligibility. In 1847, Thomson had this to say about a mathematical theory of electric charge distribution that had been developed by the Frenchman Siméon Denis Poisson (1781–1840) in 1812: "the physical interest has been almost lost in the struggle with mathematical difficulties, and the complexity of the solution has eluded that full interpretation without which *the mind cannot be satisfied* in any analytical operations having for their object the investigation or expression of truth in natural philosophy."[6]

For Thomson, truth and mechanical explanation went hand in hand. But what, precisely, did he mean by "mechanical" explanation? Isaac Newton's mechanics included the inverse-square law for gravity, but gravitational attraction bore little apparent relationship to the forces that could be employed in the fabrication of a steam engine or a clock. Although weight had always played an important role in engineering, it was simply a given that could be applied to understanding a machine's functioning; no engineer bothered to explain *why* bodies are heavy. But if mechanical intelligibility were to be the standard for explanations of nature, then weight—gravity— would also have to be dealt with; *all* forces would need to be understood as mechanical effects. Forces in a machine were exerted by one part pushing against another, and that made sense. So sense would be made of apparent deviations from that model, such as electrical and magnetic forces, or gravity and the behavior of light, by the postulation of some material substrate to transmit them—in other words, an aether.

In 1856 Thomson published a paper that exemplified his mature approach to such questions. The paper concerned an experimental finding of Faraday's from 1846 known as magneto-optic rotation. This was a phenomenon that related the transmission of light to the question of the physical reality of magnetic lines of force. Faraday had wanted to show that magnetic effects were not caused by some unknown configuration of the matter (whether solid, liquid, or gaseous) through which the magnetic lines seemed to pass. He wanted to show instead that the lines had a real existence in space, independent of the material that revealed them by its behavior in their presence. By reflecting light from a sheet of glass, he produced a beam of polarized light; that is, light in which the transverse waves all move in the same plane rather than being a jumble of waves in all possible planes. Faraday allowed this beam to pass through a block of glass, while an electromagnet was arranged so that its north and south poles were in line with the direction of the light ray. Faraday found that the plane of polarization of the light was rotated by the light's passage through the glass when the magnet was in operation (the same result occurred with an ordinary permanent magnet). The rotation increased when the strength of the magnetism was increased (fig. 5.3 shows another of Faraday's experimental demonstrations of the effect).

Faraday interpreted his results as showing a direct, unmediated effect of the magnetic lines of force on the beam of light—even though the matter of the glass was necessary to produce the effect. He argued that whatever the arrangement of the particles in the glass might happen to be, it could not be the cause of the rotation. If the magnetism had an effect on the light solely by virtue of creating a particular asymmetrical structure of atoms in the glass (say, by action at a distance), that structure would always make the plane of polarization turn in the same direction relative to the glass. However, Faraday's results showed that the rotation of the light's plane of polarization was the same regardless of whether the beam traversed the glass in one direction parallel to the supposed magnetic lines or in the opposite direction. He supported his interpretation still further by spinning the glass and observing that its own motion left the phenomenon unaffected. As Faraday wrote, "I could not perceive that this power [to rotate the plane of polarization] was affected by any degree of motion which I was able to communicate to the diamagnetic [i.e., the transparent medium through which the magnetic lines passed], whilst jointly subject to the action of the magnetism and the light."[7]

Thomson wanted to find a way of making sense of this phenomenon in terms of a mechanical picture of light transmission. In the 1840s he had conceived the transmission of polarized light as occurring in an aetherial medium that was, of course, an elastic solid. But

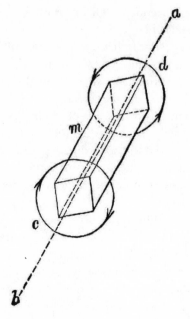

5.3. *Another demonstration of magneto-optic rotation as reported by Faraday in his "Experi-mental Researches in Electricity" in 1845. His diagram depicts a glass box filled with turpen-tine (*m*) through which a polarized ray of light (*ab*) passes. Turpentine ordinarily rotates the plane of polarization of light; an observer at* b *when the light shines from* a *to* b *will see the plane rotated in a clockwise sense, and an observer at* a *when the light is passed from* b *to* a *will also perceive a clockwise rotation (as indicated by the respective circles* c *and* d*). But when a magnetic field is set up such that its lines run through the length of* m, *the observed rotation as seen from one direction is increased, while the rotation from the other direction is reversed: the magnetic field either adds to or subtracts from the usual amount of rotation depending on the light ray's direction.*

accounting for the rotation created by Faraday's magnetic lines re-quired the addition of another mechanical feature: something in the aether must be rotating so as to impart rotation to the light waves. Accordingly, Thomson postulated that the magnetic lines of force represented rotational strain; in effect, they were twists in the aether. Thomson described this model as a "mechanico-cinematical" representation of magnetic and associated electrical forces.[8] He did not claim that this "representation" was a literal account of how these forces really worked, but that it provided a picture of them that could be described in the familiar language of mechanics and motion— and as such, served to render them in some way more intelligible.

The mechanical approach to magnetic and electric fields (to use Faraday's term) was not fully developed until another element was added—the new concept of *energy*. The 1840s and '50s saw the rapid emergence of the fundamentals of thermodynamics, in which British physicists played a central role. The steam engine lay at the foundations of nineteenth-century energy physics, since it was the central, as well as economically important, example of the use of heat to produce mechanical work. The study of heat engines by the French engineer Sadi Carnot (1796–1832) in the 1820s, joined with the crucial experimental investigations of the Manchester brewer James Prescott Joule (1818–89) in the 1840s, had contributed to the emergence in the 1850s of a new science of energy. Talking about energy and its conservation began to be the favored way of making sense of many things that had previously been understood using the Newtonian concept of force, and added a crucial new analytical dimension to the mechanical pictures of nature that were increasingly employed by the British physicists.

The new concept of energy amounted to a grand extension of the old engineer's concept of "work," which was measured (or defined) as the weight of a body multiplied by the height through which it was raised or lowered. "Energy" named something that was conserved when a process produced a new physical effect or result from the action of something old: heat producing mechanical work (raising weights) through the use of a steam engine, or electric current producing heat in a wire, or chemical phenomena creating electricity. Joule showed, through precise quantitative experiments, ways of measuring the amount of heat that could be produced from mechanical work. In one case, he used a descending weight, which gave a measure of mechanical work, to drive a generator (an invention of Faraday's) that produced electricity; the electricity in turn heated a measured amount of water. The rise in temperature of the water could then be treated as equivalent to the mechanical work expended by the descending weight, to give what Joule called the "mechanical equivalent of heat." The subsequent generalization of such equivalences by Hermann von Helmholtz (1821–94) in Germany, together with additional theoretical work by Thomson and other Britons, and the mathematical formalizations of the Swiss physicist Rudolf Clausius (1822–88), yielded by the mid-1850s the field of energetics, or thermodynamics. Energy was a very abstract concept, with no pure form that could be put in a bottle on the laboratory shelf, but its use as an accounting procedure for tracking changes in physical phenomena quickly made it indispensable. British physicists like Thomson integrated it into their own understandings of what mechanical explanations of physical phenomena involved—after all, what could be more mechanical than a steam engine?

II. Dynamical Illustration

In his paper of 1855, described above, Faraday wrote that "gravity cannot be carried on to produce a distant effect except by some interposed agent fulfilling the conditions of a physical line of force."[9] His understanding of distance forces, expressed in talk of "lines of force," related to his tangible experimental researches. For Faraday, lines of force were things that he manipulated in his experimental work; they were not abstract theoretical postulates. Just as he handled wires and magnets, so too he handled force lines, which existed for him almost tangibly, in and around his apparatus. That he had difficulty in explaining what these lines really were was a matter of less pressing importance to him; the lines seemed to exist because they had real effects. Their intelligibility resided for Faraday in their immediacy—in the fact that he could manipulate them. However, many of his fellow physicists, such as Thomson and Maxwell, tended to be theoreticians with a mathematical rather than a qualitative, experimental bent. For them, intelligibility resulted from being able not just to manipulate, but to account for field lines in a mathematical-mechanical language. This language would ground physical phenomena in what they took to be more fundamental features of reality: Newton's laws of motion and the law of conservation of energy, applied to inert matter.

Thomson used a significant term to denote imaginary physical models designed to account for natural phenomena: "dynamical illustrations." The point of a dynamical illustration was to show that a given phenomenon (such as magneto-optic rotation) *could* be accounted for by an imagined material aether that obeyed the basic laws (now including those of energy) governing ordinary mechanical systems. Whether nature really was like that—whether the particular model was *just like* the natural phenomenon—might be an ultimately undecidable question. The important feature of such illustrations was that they showed a consistency between the phenomenon's actual existence and the possibilities inherent in "dynamical" principles: if it was possible to account for a phenomenon in dynamical terms, then that phenomenon had been shown to be dynamically intelligible. Thomson wrote:

> The explanation of all phenomena of electro-magnetic attraction or repulsion, and of electro-magnetic induction, is to be looked for simply in the inertia and pressure of the matter of which the motions constitute heat [an important form of energy]. Whether this matter is or is not electricity, whether it is a continuous fluid interpermeating the spaces between molecular nuclei, or is itself molecularly grouped; or whether all matter is continuous, and molecular heterogeneousness consists in finite vortical or other relative motions of contiguous parts of a body; it is impossible to decide, and perhaps in vain to speculate, in the present state of science.[10]

Thomson made these remarks concerning his 1856 work on Faraday's magneto-optic rotation. They represent his views at that time of what a satisfactory physical explanation should include. Faraday's lines of force needed to be expressed in the terms of mechanical action in a medium.

Thomson's attempt on this problem of magneto-optic rotation now involved an aether that contained molecular vortices, actually spinning whirlpools or tornadoes in a fluid medium, instead of rotational strain in an elastic solid. These vortices twisted the plane of polarization of light-waves traveling parallel to their axes. Thomson's point was not to suggest anything very precise about the real mechanical structure of the aether, but simply to indicate the in-principle possibility of explaining this phenomenon mechanically, thereby rendering it intelligible: rotation of light would make sense if there were some kind of rotation in the medium that transmitted it.

Thomson's ambitions for his mechanical conception of natural philosophy extended even further than understanding electromagnetic phenomena. In 1858, Helmholtz published a study of vortex motion, including examples such as smoke rings. Helmholtz showed that such motions in a fluid medium were inherently stable; that, in the absence of frictional dissipation of energy, smoke ring vortices would not spontaneously fade into the surrounding medium, but would last indefinitely. In 1862 Helmholtz visited Thomson in Glasgow; subsequently P. G. Tait, Professor of Natural Philosophy at Edinburgh, showed Thomson how to produce smoke rings easily by taking a box with a circular hole in one side, filling it with smoke, and then tapping it. In 1867 Thomson used these ideas of the stability of vortices to write his paper "On Vortex Atoms." Earlier that same year, he had written to Helmholtz on his ideas:

If there is a perfect fluid all through space, constituting the substance of all matter, a vortex-ring would be as permanent as the solid hard atoms assumed by Lucretius [an ancient Roman atomist] and his followers (and predecessors) to account for the permanent properties of bodies (as gold, lead, etc.) and the differences of their characters. . . . [A] long chain of vortex-rings, or three rings, each running through each of the

others, would give each very characteristic reactions upon other such kinetic atoms.[11]

This was a very ambitious vision for physics, one in which all kinds
of physical and chemical properties of matter would be accounted
for in terms of mechanical action in a fluid aether.

In a sense, Thomson wanted to develop a version of what is now
sometimes called a "theory of everything." In 1870 he wrote: "Is action at a distance a reality, or is gravitation to be explained, as we
now believe magnetic and electric forces must be, by action of intervening matter?"[12] His implied answer was clear: only the latter was
acceptable. The details of such a program by now rested on a careful exposition of dynamical explanation as a general approach to
physics. This took the form of a book that he had written with Tait;
their *Treatise on Natural Philosophy* (1867) was intended as a general introduction to mechanics and the physical sciences, although during
the several years of its composition it became less elementary than
was originally intended. In the end, only the first volume of the pro-
jected work ever appeared.

The book's preface shows how its authors saw the relationship between, on the one hand, dynamics and mechanics, and, on the other,
a broad conception of natural philosophy.

The term Natural Philosophy was used by NEWTON, and is still used in
British Universities, to denote the investigation of laws in the material
world, and the deduction of results not directly observed. Observation,
classification, and description of phenomena necessarily precede Natural Philosophy in every department of natural science. The earlier
stage is, in some branches, commonly called Natural History; and it
might with equal propriety be so called in all others.[13]

The *Treatise* presents the study of motion and forces as fundamental
to natural philosophy, and displays an aversion to action at a distance.
There is little doubt that the limited conceptual canvas used in

Thomson and Tait's preface represented, for them, the proper limit on what ingredients could be said to be truly intelligible—so obviously intelligible, in fact, that the authors made little fuss about the question. They treated natural philosophy as if it were a fundamentally straightforward enterprise based on observation and experiment, with "laws" of nature as its ultimate constituents. The virtues of the dynamical approach were, after all, reckoned to be self-evident.

III. Maxwell and the Aether: Mechanics as Understanding

Maxwell saw his job as a physicist in very much the same light as Thomson. But his devotion to mechanical-dynamical accounts of physical phenomena was even less concrete than Thomson's. Thomson, in implicating energy in his "dynamical illustrations," tended to treat mechanical energy—strains in a medium, or the motion of a mass—as fundamental. Maxwell, by contrast, tended to shy away from the conclusion that all processes in nature were necessarily strictly mechanical. In an 1875 article on "Attraction" in the *Encyclopaedia Britannica*, Maxwell wrote that the physical problem of gravitational attraction between bodies really amounted to asking "Why does the energy of the system increase when the distance increases?"[14] This reference to potential energy (a term coined by another British physicist, W. J. Macquorn Rankine) did not in itself presuppose a mechanical interpretation of energy. But Faraday's discussion of magnetic lines of force had focused on the medium between bodies, and Maxwell accepted that "Thomson afterwards proved, by strict dynamical reasoning, that the transmission of magnetic force is associated with a rotatory motion of the small parts of the medium."[15] Maxwell found various problems in existing accounts of gravitational attraction that proposed such models, but he accepted that whatever was really going on, a medium was somehow involved.

Maxwell, just as much as Thomson or Rankine, made sense of electromagnetic phenomena by devising precise mechanical models. The purpose of such models was, as with Thomson's vortex proposal, to show how the phenomena could be made consistent with mechanical principles, and hence intelligible. Maxwell's first paper on the subject, "On Faraday's Lines of Force," appeared in 1856. Although the paper failed to develop a full mechanical interpretation of the lines (an inadequacy that Maxwell later attempted to rectify), it proceeded by setting up a physical analogy between Faraday's lines of force and the motion of an incompressible fluid flowing through tubes; its only goal was to create a mathematical formalism to describe the lines. But even though Maxwell did not claim in this paper to show in mechanical terms what lines of force really were, he did regard them, as did Faraday, as having a real, independent physical existence; they were not just computational fictions.

In 1861 and '62, Maxwell published what he considered to be, in contrast to the earlier paper, a proper physical treatment of electromagnetism. This four-part paper, "On Physical Lines of Force," regarded the equations that he had produced in 1856 as simply redescriptions of Faraday's own experimentally validated ideas about the lines. Hence, despite the mechanical analogies that Maxwell had used in developing them, the equations' validity did not rely on the details of any particular mechanical model. At the same time, Maxwell believed the supposed rotatory motions in the aether to be real, and now wanted to account for them in terms of forces capable of producing them.

> My object in this paper is to clear the way for speculation in this direction [i.e., lines of force as physical states or actions of a medium] by investigating the mechanical results of certain states of tension and motion in a medium, and comparing these with the observed phenomena of magnetism and electricity. By pointing out the mechanical consequences of such hypotheses, I hope to be of some use to those who consider the phenomena as due to the action of a medium, but are in doubt as to the relation of this hypothesis to the experimental laws already established, which have generally been expressed in the language of other hypotheses.[16]

Maxwell now proceeded to design a mechanical model of the structure of the aether that would be consistent with the electromagnetic
phenomena (fig. 5.4).

He first represented magnetic lines by rotating tubes, or vortices.
The direction and rate of rotation of the vortices corresponded to
the direction and strength of the magnetic field in that region of
space; the vortices were all packed together like a bundle of un-
cooked spaghetti, with no variation in the density of their packing.
In asking himself how to fit the empirically known relationship between magnetism and electric currents into this picture, Maxwell
then appealed to an additional consideration: the limits of his own
understanding.

> I have found great difficulty in conceiving of the existence of vortices in
> a medium, side by side, revolving in the same direction about parallel
> axes. The contiguous portions of consecutive vortices must be moving
> in opposite directions; and it is difficult to understand how the motion of
> one part of the medium can coexist with, and even produce, an oppo-
> site motion of a part in contact with it.
>
> The only conception which has at all aided me in conceiving of this
> kind of motion is that of the vortices being separated by a layer of par-
> ticles, revolving each on its own axis in the opposite direction to that of
> the vortices, so that the contiguous surfaces of the particles and of the
> vortices have the same motion.[17]

He subsequently called these particles "idle wheels." According to
Maxwell's own account, their presence in his mechanical model of
the aether was not necessitated by the mere fact of a mechanical
setup; instead, he used a kind of autobiographical account of his
own "great difficulty in conceiving" of a workable solution to the
mechanical problem, a problem that he found "difficult to understand." His justification for introducing his solution was that it was
the "only conception" he could manage. The language of intelligibility, that is, played a leading role in Maxwell's argument.

Later, in summarizing his assumption, Maxwell acknowledged

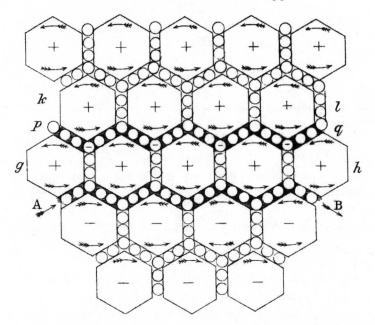

5.4. Maxwell's diagram, "On Physical Lines of Force" (1861–2), of a cross-section of his electromagnetic aether, by means of which he developed a mathematical theory of Faraday's lines of force and determined that this kind of mechanical aether would convey waves at roughly the speed of light. There are errors in the arrows on the vortices below the line of current AB; they should all designate clockwise rotation, opposite to that above the line.

the implausible artificiality of these idle-wheel particles—which, nonetheless, he had been able to use to represent, when in motion, electrical current.

> The conception of a particle having its motion connected with that of a vortex by perfect rolling contact may appear somewhat awkward. I do not bring it forward as a mode of connexion existing in nature, or even as that which I would willingly assent to as an electrical hypothesis. It is, however, a mode of connexion which is mechanically conceivable, and easily investigated, and it serves to bring out the actual mechanical con-

nexions [i.e., those forces and motions seen in experimental situations] between the known electro-magnetic phenomena.[18]

Maxwell's chief argument for introducing this feature of the model is, then, that it is "mechanically conceivable." In that sense, it was like the whole model: it was not something claimed as "existing in nature," but it did make sense of things.

The climax of Maxwell's paper comes in part 3. There, he used his model of an elastic, mechanical aether (including, crucially, the idle-wheel particles) to determine the amount of elasticity that a medium of that sort should possess if it were the cause of the electromagnetic forces measured by experiment. The result showed him that waves in such a medium would travel through it at a speed very close to the measured speed of light in a vacuum, and hence that light likely was itself an electromagnetic disturbance.

Although Maxwell routinely referred to his model in this paper as "hypothetical," one aspect of it that was clearly *not* meant to be hypothetical was its assumption that electromagnetic forces and wave disturbances took place in a medium of some kind. The fact that Maxwell also viewed this medium as specifically mechanical reflects his view of the special intelligibility of mechanical (dynamical) concepts. These concepts had the further advantage of correlating nicely with the mechanical phenomena of forces and motions that were the results of experiments on electromagnetism. Maxwell's subsequent paper of 1864, "A Dynamical Theory of the Electromagnetic Field," attempted to dispense with the details of his aether model, although it still took for granted that a material aether existed to sustain and transmit forces: "the parts of this medium must be so connected," he explained, "that the motion of one part depends in some way on the motion of the rest; and at the same time these connexions must be capable of a certain kind of elastic yielding, since the communication of motion is not instantaneous, but occupies time."[19] In developing particular mechanical inferences from these starting points, he described what he came up with as a "dynamical illustration" of these fundamental points, using the expression in the sense previously established by Thomson.[20] But Maxwell was careful not to claim too much reality for the mechanical terms that he used when talking about the aether: "I wish merely to direct the mind of the reader to mechanical phenomena which will assist him in understanding the electrical ones."[21] The professed goal of Maxwell's enterprise was the creation of *understanding*.

In his *Treatise on Electricity and Magnetism* of 1873, Maxwell, like Faraday before him, invoked Newton as an authority on the implausibility of genuine action at a distance and reaffirmed his own view that electromagnetic action is a property of "the medium in which the propagation takes place."[22] His remarks were directed against certain German physicists who persisted in using the idea of action at a distance in explaining these same phenomena. However, Maxwell also refused to commit himself to the necessary truth of his dynamical characterization of the aether, instead preferring to stress the value of such an approach for the "understanding" of the situation; this careful nuance distinguished him to some degree from other British physicists, such as Thomson. In 1878, in an article in the journal *Nature*, Maxwell even suggested that another colleague, Rankine, who was much less cautious in these matters, owed his own predilections to his habits as an engineer:

> Whatever he imagined about molecular vortices was so clearly imaged in his mind's eye that he, as a practical engineer, could see how it would work. However intricate, therefore, the machinery might be which he imagined to exist in the minute parts of bodies, there was no danger of his going on to explain natural phenomena by any mode of action of this machinery which was not consistent with the general laws of mechanism. . . . Being an accomplished engineer, he succeeded in specifying a particular arrangement of mechanism competent to do the work.[23]

Action in a medium made sense to Maxwell, as did dynamical accounts of its structure. Even though he was less sure than others that mechanics exhausted the contents of the physical world, he could say little about any alternatives. And while Maxwell's mathematical formalisms certainly seemed to work instrumentally, they were never enough to satisfy his understanding.

James Clerk Maxwell died in 1879, but after him came numerous physicists eager to pursue the dynamical program shared by his *Treatise* (which itself proposed specific, though still conjectural, mechanical models for the aether) and Thomson and Tait's *Treatise on Natural Philosophy*. There followed continued attempts to devise more satisfactory models of an aether that could account for the forces of electricity and magnetism. In 1885, for example, G. F. Fitzgerald (1857–1901) presented an aether model containing spinning wheels, similar to Maxwell's model but with a simplified set of mechanical interlinkages. The model was capable, among other things, of representing the effect of differential electrical charges, but only by using rubber bands to take on greater and lesser degrees of strain—perhaps a rather overliteral representation that removed some of the mechanical simplicity of Maxwell's attempts. In the by-now usual way, Fitzgerald regarded his models as analogies to the true state of affairs, rather than strict representations. In 1893, a preeminent aether theorist, Fitzgerald's fellow Irishman Joseph Larmor (1857–1942), presented to the Royal Society a dynamical theory of the aether that incorporated vortex atoms of the kind previously proposed by Thomson (now Lord Kelvin, and president of the Royal Society). Including ordinary matter in his picture of an electromagnetic aether left Larmor with only gravity left to account for—but that proved to be too much.

IV. The Instrumentality of Electromagnetism

The natural phenomena that formed the focus of these British physicists' work were all created in the laboratory, using specially contrived apparatus. Only careful manipulation could yield the often delicate experimental results that informed their attempts at understanding electromagnetism. In other words, the object of their studies was bound up from the start with apparatus and the deliberate production of physical effects; they were always concerned with making and directing phenomena. When such deliberate contrivance produced results that showed a potential for practical use, therefore, some of the physicists, particularly those with a background or interest in engineering, were ready to turn their practical talents to instrumental, economically valuable, enterprises.

The central area of practical work in which these physicists involved themselves was that of cable telegraphy. The transmission of signals over long distances by means of electrical cables radically changed communications worldwide, and the far-flung British Empire was to benefit enormously from its development. In the forefront of such work were, among others, Faraday and Thomson. Submarine telegraph cables were especially significant, both because of their value in communicating between places separated by expanses of water and because Faraday and Thomson showed that such cables were particularly suited to long-distance transmission. In 1854 Faraday reported that underwater lines, like underground cables, were capable of carrying unblurred signals over a longer distance than were land lines strung on telegraph poles. Faraday's work was done at the invitation of the Electric Telegraph Company, which afforded him long telegraphic cables with which to experiment (telegraphy having previously been the province of practical electricians rather than physicists). Faraday was clearly pleased not only by the recognition of his expertise in such a practical area, but also by the possibility of showing the value of a physicist's knowledge: "when the discoveries of philosophers and their results are put into practice, new facts and new results are daily elicited."[24]

The principal scientific significance of Faraday's findings lay in his conceptions of electromagnetic phenomena as occurring in the space around an electrified or current-carrying wire, rather than being propagated by action at a distance out from the wire itself. The fact that the material surrounding the cable made such a difference to the sharpness of the signal received at the far end meant, thought Faraday, that his ideas were right: it was the field existing in the space around the wire that was important. Water, as in the case of submarine cables, or damp earth in the case of underground ones, sustained these fields better than the air surrounding lines strung along poles. In addition, something could be done to improve matters by paying more attention to the insulating material sheathing the wire in the cable itself.

William Thomson also became centrally involved in issues of telegraphy, and especially in the grand project of a transatlantic telegraph cable. Thomson came up with the basic equation governing the change in current and voltage at the end of a submarine cable, which served to predict the amount of retardation, or blurring, of a signal sent along it. After a couple of false starts, the first successful transatlantic cable was laid in 1866, rapidly followed by others stretching around the world (fig. 5.5). British companies dominated the market, and British economic as well as political might was a principal beneficiary of the new technology. But another major beneficiary was Faraday's field approach to electromagnetism: the use of his ideas in the service of telegraphy promoted their use by other British physicists, and encouraged the further development of field theory by his compatriots. The common British opposition to action at a distance, and the corresponding fondness for field conceptions, had a professional as well as a cognitive component. Thomson commented on the question in 1871:

> This leads me to remark how much science, even in its most lofty speculations, gains in return for benefits conferred by its application to promote the social and material welfare of man. Those who perilled and lost their money in the original Atlantic Telegraph were impelled and supported by a sense of the grandeur of their enterprise, and of the world-wide benefits which must flow from its success; they were at the same time not unmoved by the beauty of the scientific problem directly presented to them; but they little thought that it was to be immediately, through their work, that the scientific world was to be instructed in a long-neglected and discredited fundamental electric discovery of Faraday's, or that, again, when the assistance of the British Association was invoked to supply their electricians with methods for absolute measure-

5.5. W. H. Russell's The Atlantic Telegraph *(London, 1866; illustrations by Robert Dudley) depicts the machinery on board the* Great Eastern *for paying out the transatlantic cable. Its many wheels remind us of the central motif of Maxwell's aether model.*

ment . . . they were laying the foundations for accurate electric measurement in every scientific laboratory in the world, and initiating a train of investigation which now sends up branches into the loftiest regions and subtlest ether of natural philosophy.[25]

The "subtlest ether of natural philosophy" benefited from such practical concerns because natural philosophy drew its points of reference from other, wider areas of life; what made sense in the everyday Victorian world of commerce and industry rendered associated concepts in natural philosophy more familiar, and hence more acceptable.

A different illustration comes from a draft passage that Thomson wrote for the *Treatise on Natural Philosophy*. While speaking of the Newtonian concept of "action," he wished to translate it into the language of work: "the 'action' of a 'labouring force' at any instant is its *rate of performing work*. . . . Thus the *action of a steam engine* is a very intelligible phrase, and may be used without any vagueness to express the energy with which the engine is working."[26] An article by Maxwell on the atom in the mid-1870s drew on related aspects of Victorian industrialism to make a natural-theological point. Following John Herschel, Maxwell observed that the absolute physical identity of all atoms of the same kind was like the identity of similar manufactured articles. Mass production (perhaps powered by steam engines) implied deliberative intelligence, and since atoms were morally the equivalent of identical factory products, they too must be the result of intelligent design.[27] In 1802, William Paley's God had been a craftsman; now Maxwell's was a Victorian industrial manufacturer.

British physicists towards the close of Queen Victoria's reign lived in a world dominated by British economic and political might. It was a world transformed by new means of communication and powered by industrial machinery. In understanding this world, the physicists, acting as natural philosophers, tried to make nature into a vast mechanical engine that only their expertise could master. At the same time, now acting as instrumental scientists, they were able to use this expertise to participate in an imperial enterprise that they helped to make natural. Like the law of supply and demand, it just seemed to make sense.

CHAPTER SIX

How to Understand Nature?
Einstein, Bohr, and the Quantum Universe

Quantum mechanics is perhaps the most obvious example from the twentieth century of controversy over scientific intelligibility. Emerging from work at the close of the nineteenth century on light radiation and heat, its techniques soon raised questions about philosophical coherence and understanding.

The origins of quantum mechanics in the 1920s are a well-studied subject in the history of science. One of the reasons for the great interest has been the sense that a fundamental change occurred at that time in the received physical worldview; as a result, the universe came to appear as a strange, counterintuitive realm of imprecision and unpredictability, in which paradoxes became the norm and causality ceased to hold sway. In fact, intelligibility was centrally at stake in the advent of quantum mechanics, because considerations of what did and did not make sense became crucial to physicists' judgments about whether quantum mechanics was an acceptable theory of nature or just a practical tool for computation—and, if the latter, whether pure instrumentality was a good enough ideal for science.

I. The Quantum of Action

The idea of the quantum of energy was introduced as a kind of energetic atomism, in which energy could not be divided into as small an amount as you liked, but existed as discrete, albeit tiny, packets. In its origins, this concept was thoroughly practical. The German physicist Max Planck (1858–1947) found in 1900 that it allowed him to create a neat mathematical model of a vexing phenomenon, called black-body (or cavity) radiation. This is the kind of radiation that is emitted by a warm body but is entirely absorbed again, continuously, by that same body. As a result, none of the radiation escapes; the body appears to be black, since it gives off no radiation to the outside world. An example of such a thing would be an enclosed cavity, the internal surface of which continuously emits and absorbs radiation of all frequencies, including radio waves, visible light, and X-rays, across the entire electromagnetic spectrum.

Planck worked with careful empirical measurements of the actual energy distribution, across the different frequencies, found in black-body radiation. His goal was to develop a mathematical account of the distribution that would make sense of it theoretically. He imagined the distribution curve in terms of an array of tiny oscillators, in which each successive oscillator vibrated at a slightly higher frequency than the one before it—like a succession of musical strings tuned to progressively higher notes. Some oscillators would be vibrating more energetically than others, and the amount of energy contained in each oscillator, with its characteristic frequency, would, when added to those of the others, represent the total radiant energy of the black body at a given temperature.

Eventually, Planck found that he could get what seemed to be a good approximation to the shapes of the measured curves for different temperatures only if he assumed that the total quantity of energy in each case was distributed among the oscillators in a particular way. It did *not* work if the energy levels in the oscillators were allowed to vary continuously. Instead, the energy levels needed to be restricted to particular values that could change only in jumps, from one energy level to the next. This assumption in Planck's model violated the usual way of thinking about energy; energy, like weight, had previously been imagined as a quantity that could vary smoothly, increasing or decreasing gradually, rather than being confined to stepwise change. But doing things in Planck's way fitted the empirical data on black-body radiation better.

Planck called these discrete energy changes "quanta" of energy. He assumed that their necessity in the theory would soon be explained, and that the explanation would not require energy itself to exist only in bundles, or packets, of particular, restricted sizes. He failed to come up with such an explanation himself, however, and very soon other areas of physics began to find a use for Planck's idea. The most famous example is due to Albert Einstein (1879–1955), in a paper of 1905 on the so-called photoelectric effect. Light falling on a negatively charged metallic surface served to discharge it, creating a "photocurrent"—in effect, the light promoted the release of negatively charged electrons. Evidently the light provided energy to the electrons, enabling them to escape from the metal. But the crucial feature, for Einstein, was the fact that the intensity of the light was a significant variable only in determining the *size* of the photocurrent (the rate of release of the electrons); the strength of that current, as determined by the countervoltage that had to be applied to stop it, was dependant solely on the *frequency* of the light. That is, the amount of energy acquired by an electron from the light was due to the light's frequency and not to its intensity; the intensity determined only how many electrons acquired that amount of energy.

Einstein used this result to argue that light itself should not be understood simply as wave motion; instead, light should also be understood in terms of light *quanta*, massless packets of energy that were later to be called "photons." The actual size of the quantum of energy associated with each photon was proportional to the light's frequency. Einstein suggested that each released electron from the metallic surface acquired the requisite energy from an individual photon. So the energy of the released electrons, being derived from the energy of the photons, was, like the latter, dependent on the frequency of the light that fell on the metal. Increasing the light's intensity simply meant that there were more photons striking the metal, and hence more electrons produced; the energy provided by each photon was unaffected. The energy of a single photon, or quantum of radiation, was related to the radiation's frequency according to the expression $E = h\nu$, E being the energy, ν the frequency, and h a universal constant first derived by Planck in his work of 1900 and therefore known as Planck's constant.

The infiltration of the quantum of energy into microphysics during the first ten or fifteen years of its life was extensive. In 1913 the Danish physicist Niels Bohr (1885–1962) opened up a new and enormously fruitful arena for the use of the concept: atomic spectra. It was this work that led, over the next decade or so, to the emergence of a field known as quantum physics, from which a new "quantum mechanics" emerged in the 1920s.

Bohr's idea in 1913 was even more speculative and problematic than Planck's initial introduction of the quantum. Planck had used quanta as a kind of mathematical trick to develop a model that accorded well with experimental data but that was of obscure physical significance. Bohr adopted Planck's quantum in a similar spirit: it was, he wrote, "in obvious contrast to the ordinary ideas of electrodynamics, but appears to be necessary in order to account for experimental facts."[1] So, without worrying about its physical status, Bohr combined the idea with a speculative picture of the internal structure of atoms. He began with Ernest Rutherford's (1871–1937) famous "solar system" model of the atom, wherein negatively charged electrons orbit around the central positively charged nucleus like planets around the sun, but attracted electrically instead of gravitationally. The size of an electron's orbit corresponded to the orbit's energy, and Bohr postulated that the only orbits permitted to such electrons were those whose energy corresponded to discrete, quantized levels—not just any amount of energy, or orbit, was allowed by his model. There were levels, or steps, of orbital energy that the electrons could occupy; Bohr called them "stationary states." The only changes in orbital levels that an electron could undergo were ones in which it shifted, in a discrete jump, from one stationary state to another.

Bohr restricted his analysis to the simplest case—that of the hydrogen atom, with a single proton as its nucleus and a single electron in orbit around it. When the electron dropped from one orbital energy level to a lower one, the difference in the energy levels was lost from the system, emitted in the form of radiation. If the atom were excited to a more energetic state by the impact of radiation, the reverse would happen: the electron would jump to a higher level, representing the atom's increased energy. But all these changes were discrete, not continuous, allowing Bohr to interpret some experimental results that had been known for nearly thirty years. These results concerned the radiation (the glow) emitted by hydrogen atoms when excited by heat. When the radiation was separated out according to frequency by a spectroscope, the spectrum showed a series of discrete bright lines. There was nothing remarkable in this result itself: it had been known for most of the nineteenth century that elements, when vaporized by heat and made to glow, displayed "line spectra" of this sort. The so-called Balmer series of hydrogen lines had been found by its namesake in 1885 to fit closely a simple formula: $n \propto 1 - 4m^{-2}$, where n was the frequency of a spectral line and m was any integer from 3 to 15.[2] Remarkably, Bohr was able to derive this formula from his quantum model of the hydrogen atom (fig. 6.1).

Bohr presented this work, and its subsequent elaborations, not as straightforward confirmation of his theoretical model so much as tantalizing agreement between the theoretical predictions and experimental results. He knew that the inconsistency between his new model of the atom and the older, "classical" mechanics and electrodynamics would eventually require resolution. (In general, use of the terms "classical mechanics" and "classical physics" in this period means nothing more than the mechanics or physics that was second nature to contemporary physicists from their own prior professional training, and in which energy was always continuous, not quantized.)

When Bohr sent Rutherford the paper that first set out his new ideas (to be published in the British journal *The Philosophical Magazine*), Rutherford wrote back to express his reservations—which were of a natural-philosophical kind:

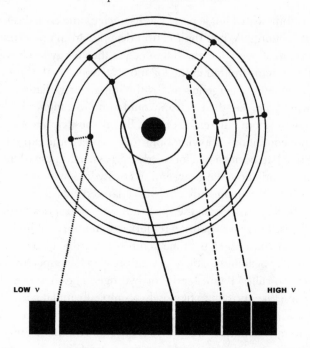

6.1. Bohr's atomic model for hydrogen and the Balmer lines in the visible part of the emission spectrum of incandescent hydrogen. Bohr's model yielded these lines on the assumption that they were due to electrons, excited to higher energy levels in the hot incandescence, randomly dropping back to the n = 2 *orbital level and releasing the particular differences in energy as light (higher frequency,* ν, *corresponds to higher energy). Other lines resulted from drops to level* n = 1, *etc.*

Your ideas as to the mode of origin of spectrum and hydrogen [*sic*] are very ingenious and seem to work out well; but the mixture of Planck's ideas with the old mechanics make [*sic*] it very difficult to form a physical idea of what is the basis of it. There appears to me one grave difficulty in your hypothesis, which I have no doubt you fully realise, namely, how does an electron decide what frequency it is going to vibrate at when it passes from one stationary state to the other? It seems to me that you would have to assume that the electron knows beforehand where it is going to stop.[3]

The leading British scientific journal *Nature* carried a report in 1913 of a meeting of the British Association for the Advancement of Science at which Bohr's theory was discussed. The eminent Dutch physicist H. A. Lorentz, the report noted, "intervened to ask how the Bohr atom was mechanically accounted for. Dr. Bohr acknowledged that this part of his theory was not complete, but the quantum theory being accepted, some sort of scheme of the kind suggested was necessary."[4]

Mathematical models using the quantum to account for the spectral lines emitted by atoms were greatly developed during the 1910s by a German physicist, Arnold Sommerfeld (1868–1951), and his students at Munich. By 1920, physicists were actively searching for something more than just applications of the quantum of action to new areas; instead, they wanted to produce what they called a "quantum mechanics," which would be a fundamental theory capable of accounting for physical behaviors at the atomic level. It had become increasingly less plausible that the quantum of action could be explained in terms of so-called classical mechanics, the kind of mechanics associated with Newton's name. Somehow, on this new view, the quantum itself would have to be made a basic part of a new mechanics. The older mechanics spoke of force, mass, and also, since the middle of the nineteenth century, energy. A quantum mechanics would also speak of such things, but the way in which they would be interrelated for atomic-level phenomena would differ from what appeared to be the case at the everyday level of ordinary matter as well as at the cosmic scale. Bohr and others assumed that a new quantum mechanics would yield results that corresponded to the results given by classical mechanics when the scale of the phenomena grew far above the atomic scale. Bohr soon elevated this assumption to what he called the "correspondence principle." The principle had the effect of constraining the kinds of quantum mechanics that could be proposed, by forbidding any observable incompatibility beyond the atomic scale between their results and those of the older classical mechanics.

The search for a quantum mechanics was therefore explicitly a natural-philosophical enterprise. It was an attempt to invent a comprehensive physical theory that would hold a fundamental status as an account of the world, rather than simply being an instrumentally useful or "applicable" theory. In penetrating to the atomic level, physicists believed themselves to be on the threshold of a new, more profound understanding of the physical universe.

II. Quantum Mechanics

Quantum mechanics came of age in 1926, with the appearance of two rival theoretical approaches. These approaches were soon accepted as being mathematically equivalent to one another, but they were couched in the terms of very different physical intuitions. One, due to Sommerfeld's former student Werner Heisenberg (1901–76), was known as "matrix mechanics," from its characteristic mathematical technique; the other, invented by the Austrian Erwin Schrödinger (1887–1961), was known as "wave mechanics," from its central physical image of matter at the atomic level. The contest between the two involved intense debate over physical fundamentals, but it also concerned the question of what scientific explanation ought to accomplish.

Schrödinger wanted to produce a quantum mechanics that would make sense of quantized energy levels on the basis of concepts that were already familiar in classical physics: waves and wave motion. In 1924, the French physicist Louis de Broglie (1892–1987) had proposed that the behavior of photons, the quantized light particles introduced by Einstein in 1905, was in fact typical of ordinary matter too (the photoelectric effect itself had by then been shown to be true of X-rays as well as visible light). Light exhibited wave properties, as if it were indeed a ripple effect in an aether, even though it also sometimes acted as if composed of discrete particles. So too, suggested de Broglie, the constituent particles of matter, such as electrons (first proposed in the 1890s as the reality behind cathode rays), also

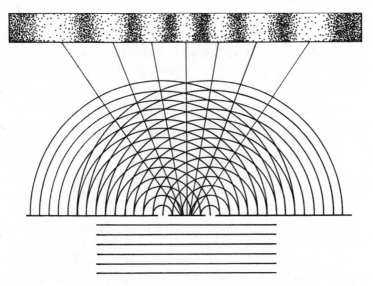

6.2. Interference fringes in the case of light, were, de Broglie suggested, also to be expected in the case of electrons. The two slits in the screen generate two sets of circular light wave emissions; where the peaks of the waves coincide, the light intensity increases, but where they are of opposite sense (one up and one down, for transverse waves), they cancel out and, when exactly equal (but opposite) in size, display total darkness.

possess wave characteristics such as interference (fig. 6.2). The relationship between a particle's mass and its associated wavelength, according to de Broglie's argument, was given by the equation $\lambda = h/mv$, where λ was the wavelength, h Planck's constant, m the particle's mass, and v its velocity of motion.

De Broglie represented his wave/particle picture as a relationship between a discrete particle and a spatially extended wave somehow associated with, or accompanying, it. Schrödinger now put forward an equally problematic but, he hoped, conceptually neater picture in his new quantum mechanics. Schrödinger imagined the electrons in an atom as being, essentially, waves. But when they were moving in free space instead of being confined to the atom, electrons seemed to have characteristics typical of particles. According to

Schrödinger's ideas, those particle characteristics were actually due to the electrons being "wave packets" rather than true, discrete particles. He imagined the packets as wave motions in particular regions of space that corresponded to the positions of the electrons. Unfortunately, Schrödinger could not account for why such wave motions did not simply dissipate, with the waves spreading out into space like ripples on a pond. Schrödinger needed his wave packets to remain localized, that is, even in free space.

In the case of an atom, Schrödinger explained why Bohr's electron "orbitals" consisted only of a restricted, quantized set of possible energy levels. A classical orbital model of the sort applicable to a solar system, after all, would have allowed orbits at any distance from the nucleus, corresponding to any intermediate energy level at all. Waves gave him the explanation that he needed, because they too could exhibit "jumps" like those of electrons between quantum levels in an atom. Imagine a vibrating string fixed between two clamped endpoints a meter apart. The string can vibrate back and forth in the form of a single half wave, that is, with a half wavelength of exactly one meter; but it can also vibrate in the form of two half waves, each with a half wavelength of fifty centimeters; or three, of a third of a meter each; and so on indefinitely (fig. 6.3).

In effect, the string's wave behavior is quantized—it cannot vibrate with half wavelengths that do not divide exactly into the distance between the clamps, and vibrating strings of this sort can indeed be made to "jump" from one set of waves to another (that is what the harmonics of musical strings are about). Certainly, the *energies* associated with these wave motions are not quantized (since they depend also on the amplitude of vibration), but Schrödinger saw the wave approach as a way of reminding physicists that a restriction to discrete values did not necessarily have to violate the continuities familiar in classical physics.

Soon, Schrödinger produced a demonstration that both versions of quantum mechanics, the matrix mechanics of Heisenberg and his own wave mechanics, were mathematically equivalent. This came as a relief to Heisenberg and his chief allies, among them Bohr and

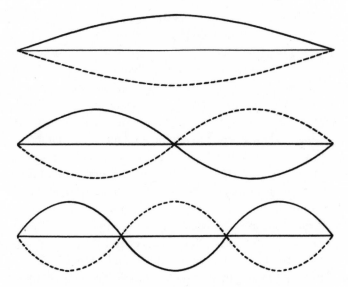

6.3. The first three harmonics of a musical string: all possible modes of vibration must contain whole numbers of half waves, here 1, 2, 3; in effect, such standing waves between two fixed points must be quantized (although not in relation to their energy content).

Wolfgang Pauli (1900–1958). This so-called Copenhagen group (named after the location of Bohr's physics institute) had mistrusted Schrödinger's theory because, among other things, it threatened to put matrix mechanics into the shade. Schrödinger's wave approach had been quickly adopted by physicists because it lent itself much more readily to the solution of specific physical problems; it was also a visualizable theory, in contrast to the abstract mathematical techniques offered by Heisenberg.

The historian and philosopher of science Mara Beller has argued that matrix mechanics was originally conceived as a kind of particle (rather than wave) theory. But in practice, and on the basis of Heisenberg's published work, matrix mechanics functioned in its first couple of years as a mathematical formalism with no clear physical meaning at all—indeed, in late 1925, when first presenting matrix mechanics, Heisenberg explicitly characterized it as being concerned only with the correlation of observable quantities rather than providing some kind of underlying physical explanatory system. He therefore rejected what he called *Anschaulichkeit*, that is, "picturability" or, in this context, a kind of physical intelligibility that relies on internally consistent, visualizable accounts of underlying processes. However, in the face of Schrödinger's challenge, Heisenberg soon changed tack and produced an interpretation of quantum behavior that talked about what really happened in atomic-level mechanical interactions. The strange results were expressed in 1927 as Heisenberg's celebrated uncertainty principle, which held that certain pairs of quantities, such as the momentum and position of a particle, cannot both be known simultaneously with absolute precision: the more precise the measurement of one, the more imprecise that of the other.

Heisenberg's 1927 paper bore the general title "Über den anschaulichen Inhalt der quantentheoretischen Kinematik und Mechanik," that is, "On the physical [or "picturable"] content of quantum-theoretical kinematics and mechanics." It began with a confident statement of what "understanding" in physics should be.

> We believe we understand the physical content of a theory when we can see its qualitative experimental consequences in all simple cases and when at the same time we have checked that the application of the theory never contains inner contradictions.[5]

After further discussion of this ideal in relation to quantum mechanics, Heisenberg continued:

> The question therefore arises whether, through a more precise analysis of these kinematic and mechanical concepts, it might be possible to clear up the contradictions evident up to now in the physical interpretations of quantum mechanics and to arrive at a physical understanding of the quantum mechanical formulas.[6]

The paper proceeded to lay out an account of quantum indeterminacy (uncertainty) that made its points by describing specific experiments and the behavior of elementary particles within them. This approach was evidently Heisenberg's way of linking the abstract mathematical theory of matrix mechanics to the *anschaulich*, intuitively graspable, reality of experiments and the facts that those experiments revealed about the natural world. By the end of the paper, however, Heisenberg's story took on a different tone:

> As the statistical character of quantum theory is so closely linked to the inexactness of all perceptions, one might be led to the presumption that behind the perceived statistical world there still hides a "real" world in which causality holds. But such speculations seem to us, to say it explicitly, fruitless and senseless. Physics ought to describe only the correlation of observations. One can express the true state of affairs better in this way: Because all experiments are subject to the laws of quantum mechanics . . . it follows that quantum mechanics establishes the final failure of causality.[7]

This passage makes it clear that Heisenberg was trying to have it both ways. At the outset of the paper, Heisenberg stressed picturability as a proper goal of physical theory, one requisite for physical understanding. He wanted to do this in order to counter the intuitive appeal to physicists of Schrödinger's wave mechanics. Later, however, Heisenberg effectively admitted that this picturability was in reality restricted to the instrumental readings made in actual laboratory-scale experiments. He could not offer, it seemed, a satisfactory picture of what was happening on the subatomic scale, a "picture" of events comparable to that proposed by Schrödinger. Nonetheless, he still wanted to try; this same 1927 paper also put forward an example of his new uncertainty relation in terms of the behavior of subatomic particles, and it was an example that, as his colleagues quickly told him, did not work.

Heisenberg attempted to clarify the practical meaning of his uncertainty principle: that a particle's position and its momentum could not both be determined exactly for the same moment of time. His idea was that the experimental requirements for measuring the one

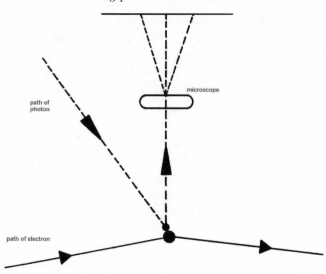

6.4. Heisenberg's flawed thought-experiment intended to show how uncertainty relations in physical measurements stem from the unavoidable characteristics of experimental situations (after Beller, Quantum Dialogue, *p.72).*

quantity necessarily precluded the experimental requirements that would allow a determination of the other. By way of illustration, he imagined an attempt to determine the exact position of an electron by illuminating it and observing it with a microscope (fig. 6.4). This device, in other words, would detect a passing electron by means of bouncing a photon off it; the photon's path would thereby be deviated into the microscope. If one knew both the photon's energy and its direction of motion immediately before the collision, as well as its energy and direction (straight into the microscope) following the collision, one would be able to calculate the electron's exact position at the moment of impact. The same experimental setup would also allow calculation of the electron's momentum at that same instant. So far, the argument follows so-called classical mechanics.

But the precision of the position measurement would depend on the wave frequency associated with the photon: the higher the frequency (that is, the shorter the wavelength), the more precisely the electron's position can be determined; the level of precision could be made arbitrarily high simply by increasing the frequency indefinitely up into the gamma ray range. Now, as we already know, the greater the frequency of a photon, the more energetic it is. The change in the electron's momentum that results from the collision is therefore itself greater when the photon has a higher frequency. And, Heisenberg says, the size of the electron's momentum change determines the precision, or rather the imprecision, with which it can be measured—the larger the change, the greater the imprecision. This is because possible momentum changes are restricted to quantized jumps, corresponding to a discontinuous array of possible energy values. Heisenberg's conclusion is that the accuracy of the position determination and the accuracy of the momentum determination for the electron are in this case inversely related: greater precision of position measurement, thanks to a higher photon frequency, corresponds to lesser precision of momentum measurement, due to the bigger energy jumps. And it is the nature of the observation itself that creates the imbalance—the effect on the electron of its being bombarded by an inquisitive photon.

Unfortunately, Heisenberg's colleagues, particularly Bohr, told him even before his paper appeared in print that this illustration was flawed. As presented, Heisenberg's *Gedankenexperiment,* or thought-experiment, fitted classical assumptions due to the very fact that it was presented as an interaction between *particles,* ignoring their wave characteristics except for speaking of the photon's "frequency." In place of discussing wave features, Heisenberg had spoken of the *discontinuity* attending the electron's change of momentum after the collision. If the collision really had been between discrete particles, then classical mechanics would have enabled precise measurement of both the electron's position and its momentum at the time of the collision, based on the colliding photon's observed behavior. Although the momentum changes would now be quantized (for whatever reason), they would still be perfectly determinable from the measurements available in this experimental situation. Bohr pointed out how this made Heisenberg's account improper as an illustration of the indeterminacy that would actually be present in this experiment. The indeterminacy would really stem, Bohr said, from the fact that the photon entering the microscope would not truly travel as a discrete particle, but would have wave properties. Consequently, a precise determination of the photon's direction would be problematic: when measuring a precise direction, which required treating the light as a particle, the light's wave features would be neglected and hence unknown; but they were just what was needed to measure the photon's energy as a way of inferring the electron's momentum prior to the collision. Correspondingly, when measuring the photon's frequency, which involved treating it as a wave phenomenon, the precise direction of its motion as a particle would be unknown (since waves spread out). The same points in general also applied to the wave characteristics of the electron.

Heisenberg acknowledged the correctness of the critique in a note appended to the published article, but he did not abandon this illustrative thought-experiment. Not only he, but also Bohr, continued to use it in popular expositions of the uncertainty principle. These expositions presented the uncertainty in the form of the famous "observer effect," whereby the act of observation itself (here, the photon collision) affects the perceived properties of the thing being observed (the electron). The example, as Beller has suggested, had the advantage for Heisenberg of being a particle interpretation of subatomic behavior, counter to Schrödinger's approach, while for Bohr, it was an easy way to illustrate what quickly became the cornerstone of his own interpretation of quantum mechanics: the role of experimental measurements in the correct application of interpretive models.

III. Making Sense of Complementarity

The difficulty by 1927 seemed to be that, in order to deal with quantum phenomena, it was necessary to think of subatomic units such as electrons as both particles *and* waves. Both conceptualizations had difficulties, but each seemed indispensable in order to explain the full range of relevant phenomena. Bohr decided that only one of the two physical pictures was needed to make sense of any particular experimental situation: sometimes the particle picture was best, at other times the wave picture, but never both at once. That is, Bohr avoided a formal contradiction between the two pictures by giving them separate jobs that never became mixed together. An experimental setup that demonstrated the wave properties of electrons was not suitable for precise measurement of those electrons' positions as particles; whereas an experimental setup suitable for measuring the position of an electron would be unable to reveal precise information about its wave properties.

Bohr explained his views in a talk at a major conference in Brussels in 1927. The talk, published in *Nature* the following year, was enormously important in creating a consensus among physicists that the conceptual problems of quantum mechanics were now, in a general way, under control. The following statement was crucial to Bohr's argument:

> Now, the quantum postulate implies that any observation of atomic phenomena will involve an interaction with the agency of observation not to be neglected. Accordingly, an independent reality in the ordinary physical sense can neither be ascribed to the phenomena nor to the agencies of observation.[8]

He explained this rather negative assertion more positively in terms of what he called a "complementarity" between the different things that could be said about atomic-level phenomena when they were examined in relation to different kinds of experiments. On the one hand, the propagation of light in time and space can be well understood when described in the terms of the wave theory. On the other hand, the so-called Compton effect needed a quite different approach. Discovered in 1922, the Compton effect showed the decrease in a photon's frequency following collision with an electron; Heisen-

berg had employed it in his uncertainty paper. The effect's explanation, in terms of such things as collision and exchange of energy,
relied on discontinuous processes that were governed by discrete
quanta of action; the same point applied to Einstein's quantum explanation of the photoelectric effect (see above). The two pictures,
wave and particle, appeared very different, and even contradictory,
but each seemed necessary so as to make sense of different experimental interactions. That is why Bohr had said that "an independent
reality in the ordinary physical sense can neither be ascribed to the
phenomena nor to the agencies of observation." Light was not to be
understood simply as waves or else simply as particles; in some circumstances of observation it had to be described as the one, in other
circumstances, as the other. Bohr's crucial point was that any particular experimental setup could be suitably accounted for only in terms
of *one* of the two complementary features—much like the reciprocal
uncertainties of measurement that Heisenberg had described.

   This notion of complementarity, or "wave-particle duality" as it
became more familiarly known, did not meet with approval from all
physicists. It accounted in formal terms for how the new quantum
mechanics could be used, in practice, to say things about the natural
world despite the apparent inconsistencies in its portrayal of that
world. But formal accounting was not enough for some; it could appear as a word trick that merely swept the problems under the rug.
Einstein found it quite unacceptable, especially in its justification of
the use of quantum mechanics to provide purely statistical accounts
of atomic processes, beyond which nothing more could be said,
rather than proper *causal* explanations. "To believe this is logically
possible without contradiction," he wrote in 1936; "but, it is so very
contrary to my scientific instinct that I cannot forego the search for
a more complete conception."[9] In fact, Bohr himself had finished his
1928 paper by referring to "the feature of irrationality characteriz-
ing the quantum postulate." But Bohr thought that he could make
sense of this apparent irrationality precisely by making the human
observer of nature part of the problem instead of part of the solution: "I hope, however," he had continued, "that the idea of com-

6.5. Natural philosophers hard at work. Left to right: an intent Bohr, a meditative (or bored) Heisenberg, and Wolfgang Pauli, 1930s. Niels Bohr Archive, courtesy AIP Emilio Segrè Visual Archives.

plementarity is suited to characterize the situation, which bears a deep-going analogy to the general difficulty in the formation of human ideas, inherent in the distinction between subject and object."[10]

A crucial component of Bohr's Copenhagen interpretation of quantum mechanics concerned the use of ordinary classical concepts in the day-to-day work of the experimental physicist. In the late 1920s and early 1930s, Bohr began to insist that the physicist's use of familiar terms such as "length," "mass," "time," and "force," which had well-defined operational meanings in the laboratory, were indispensable for the future development of physics, even though they often ran into problems when applied to the microworld of quantum mechanics. These classical concepts, Bohr maintained, derived directly from laboratory experience, and were therefore not modifiable—they were as fixed as the physical and cognitive capabilities of human beings. Since human nature could not be changed, classical concepts were indispensable in acquiring knowledge in physics; they always mediated between us and the external world—without them, we could not understand the world. The paradoxes of complementarity resulted from that fundamental truth, and were therefore themselves equally unavoidable.

Bohr's arguments amounted to an attempt to turn the logic of a particular human confection, the theory of quantum mechanics, into the logic of the world itself. But his conclusions ran into problems when they were applied to the very kinds of practical experimental situations that Bohr used to justify them. If classical concepts were necessary and proper for macro-level events such as the behavior of laboratory apparatus, but did not work at the micro-level of atomic-scale events, where should the line between the two levels be drawn? How big was big, and how small was small? Bohr offered no principled rule that could govern how to tell. But requiring a categorical difference between these two levels helped to justify the fundamental role played by acausality in dealing with atomic-level phenomena: quantum mechanics could only offer probabilities concerning, say, when the radioactive decay of an atomic nucleus might occur, instead of providing a causal account of when it will necessarily happen. Heisenberg and others used the supposed indispensability of classical concepts to argue that this quantum acausality was a feature of the world itself, rather than just a feature (or limitation) of the theory. Some events just happened. Suggestions that deterministic alternatives to quantum mechanics might still be invented, ones using new concepts different from the classical concepts that were involved in talk of waves and particles, were dismissed by Bohr and his confederates as impossible *in principle*, due to the way in which human beings relate to the world. Such alternatives could not work without destroying the basis on which we make sense of nature.

The historian of science Paul Forman has argued that this strong desire to enthrone acausality as an unavoidable implication of the new theory can be understood in the context of German cultural history in the period following the First World War. There was no logical necessity to make acausality so fundamental to the theory, says Forman, and so something else must be responsible for its strong support in the 1920s. Forman finds the answer in widespread contemporary critiques of the causality, determinism, and necessity found in the theories of modern science. These critiques were promulgated, in the wake of Germany's crushing military defeat, by figures such as Oswald Spengler in his influential two-volume work *The Decline of the West* (1918, 1922).

Forman's argument is controversial, and it spawns as many problems of historical interpretation as it purports to solve, but it draws attention to an important feature of the early history of quantum mechanics. Despite a general acceptance by scientists that the theory was instrumentally effective in making successful predictions of experimental outcomes, not everyone agreed that this success implied the reality of a fundamental acausality in nature itself. Indeed, after the Second World War, the work of a few maverick physicists, such as the American David Bohm, suggested the possibility of causal alternatives to the Copenhagen interpretation of quantum mechanics (see section VI, below), while no less a figure than Einstein consistently rejected the notion that quantum mechanics necessitated acausality in nature. Why, then, were so many physicists intent on arguing that acausality was indeed a necessary feature of the world? Forman's attempted explanation shows the extent to which issues of the perceived intelligibility, and hence scientific acceptability, of acausality in nature ultimately had to rest not on formal arguments (which could never be resolved conclusively), but on culturally rooted sensibilities.

IV. Challenges and Controversies

In the 1930s Einstein engaged Bohr in a lengthy series of controversies over the fundamental status of the new quantum mechanical accounts. Einstein's strategy rested on a natural-philosophical representation of quantum mechanics, one that Bohr himself accepted. As such, the dispute centered on the intelligibility of certain experimental situations when they were described in the terms of Bohr's version of the theory, rather than on the ability of quantum mechanics to predict quantitative results. Einstein wanted to show that, whatever probabilistic rules could be demonstrated by the formalisms of quantum mechanics, those rules were insufficient to provide a full, or complete, account of physical reality itself: nature was not fundamentally probabilistic, even if the results of quantum calculations were.

The debate centered on thought-experiments, a favorite technique of Bohr as well as Einstein. These were not a matter of performing real experiments in a laboratory, but only of imagining possible experiments and predicting their outcomes on the basis of theory. If the theory yielded a self-contradiction in its prediction of the experimental outcome, then it was defective. Einstein therefore tried to undermine Bohr's interpretation of quantum mechanics by crafting thought experiments of which Bohr's version of quantum mechanics would be unable to provide a coherent account.

The most challenging of them is known as the Einstein-Podolsky-Rosen (EPR) argument, from a coauthored paper published in 1935. It imagines two subatomic particles, initially interacting, that then begin to move apart from one another. Later, when they are far apart, a momentum measurement (for example) made on one of the particles, A, will enable a precise calculation of the value of that same quantity for the other particle, B, because the two values are linked due to the particles' earlier interaction at the common point of origin. But a difficulty arises for values of so-called conjugate pairs, that is, those pairs, such as momentum and position, that are governed by the Heisenberg uncertainty relations. At the moment when a measurement of, say, momentum is made, very precisely, for particle A, the measureable value of A's position will then be inherently uncertain. This means that the inference of particle B's momentum will also be precise, but at the same time B's position will be uncertain, because it can be inferred only on the basis of the known value of A's position, which is itself uncertain.

The paradox that Einstein wanted to identify in this situation is the following: the uncertainty in particle A's position when its momentum is measured would normally have been attributed by Bohr, or Heisenberg, to the observation itself that measured the momentum—that is, to the so-called observer effect. But this explanation cannot be applied to the uncertainty found in particle B's position, because no observation of B has been made. Why, then, Einstein wonders, should its position be uncertain? The theory does indeed give an uncertain result for the position, but why should we believe that this is anything other than a shortcoming of the theory? Why should particle B, to which nothing has been done, have an inherently fuzzy position, when, had particle A's position, instead of its momentum, been measured with precision, particle B would then have had a precise, nonfuzzy position? How, in other words, could observations made on particle A have a physical effect on a far-distant particle B? Einstein thought that this paradox proved the incoherence of Bohr's Copenhagen interpretation of quantum mechanics, according to which the probabilistic, or uncertain, features of the theory were in fact properties of the world itself. For Einstein, particle B's position and momentum must both have real, precise values out in the world, regardless of whether we can know what they are.

Bohr's answer to the challenge, also of 1935, has been described as "counterintuitive"[11] and as amounting to "no answer at all!"[12] But leaving aside the many details of Bohr's reply, one aspect of it is especially striking: it rests on assumptions about the purpose and nature of science that were very different from Einstein's. Einstein wanted physics to speak about a world that existed independently of the human observer; in effect, for the physicist to aspire to know the natural world "as God knows it." There ought, thought Einstein, to be an account of the physical world, of its contents and behaviors, that would be absolute rather than relative to human capacities and limitations. A Martian physicist, one might say, should end up with the same scientific picture of the world as its terrestrial counterpart. Bohr, by contrast, insisted on the active role of the scientist in creating knowledge of the world: experimental manipulations and the associated observation of their results could never, even in principle, be eliminated from the situation. In answering the EPR paradox, for instance, Bohr provided an analysis of the experimental setup that would be involved in actually measuring, in practice, the position or the momentum of one of the two particles. His analysis was designed to show that it was never possible to determine both the position and the momentum of one of the particles at the same time; the determinations of those values required two distinct experimental arrangements that the experimenter had to choose between. An inference to the corresponding values for the other particle could be made only for the same quantity (whether of position or momentum) as had been chosen for measurement on the first particle; the other value would simply remain unknowable—and, Bohr implied, it was therefore not meaningful to imagine that there really was a precise value there at all. Scientific truth was relative to the nature of human cognition and understanding.

In concluding his response to EPR, Bohr commented more generally:

> [T]here can be no question of any unambiguous interpretation of the symbols of quantum mechanics other than that embodied in the well-known rules which allow to predict [*sic*] the results to be obtained by a given experimental arrangement described in a totally classical way.[13]

Quantum mechanics, for Bohr, *could not* be interpreted as referring to anything other than the measurements accessible to experimental apparatus; as long as the theory made self-consistent and empirically confirmed predictions of experimental outcomes, it was true, and to speak of anything outside the realm of what was in principle experimentally accessible was nonsensical. His position was not simply a judgment on the capabilities of a fallible theory; Bohr wanted quantum mechanics to be seen as a necessary implication of the way the world works. Rather than being a mere theory, then, quantum mechanics was to be seen as a *discovery*. And what quantum mechanics could not know, human beings could not know.

The final sentence of Bohr's paper made this last point entirely clear, and it included a reference to Einstein's own work on relativity as a means of suggesting that Einstein himself ought to know better:

> [T]his new feature of natural philosophy means a radical revision of our attitude as regards physical reality, which may be paralleled with the fundamental modification of all ideas regarding the absolute character of physical phenomena brought about by the general theory of relativity.[14]

Bohr's use of the term "natural philosophy" in this passage implies that he saw his version of quantum mechanics as representing the way the world really is, not simply as predicting experimental outcomes in the laboratory. In effect, he wanted to elevate the instrumental (here, predictive) effectiveness of quantum mechanics to the status of a kind of surrogate for intelligibility itself—just as much a hallmark of successful natural philosophy as intelligibility had often been in the past. And that was the ambition that Einstein wanted to thwart.

V. Schrödinger's Cat

Schrödinger, the inventor of wave mechanics (the original alternative to Heisenberg's matrix mechanics), was also determined to show that there was something lacking in quantum mechanics and that the Copenhagen interpretation was insufficient. He wrote a paper in 1935, spurred on by the recently published EPR paradox, that attempted to refine what he saw as the central point at issue. Schrödinger's paper considered the matter in relation to the use of models in physics—such theoretical pictures as the nineteenth-century kinetic theory of gases, which imagines a gas as being composed of molecules flying about in space and interacting through collisions. Using a model like that, one could predict certain features of the actual situation, assuming that the model corresponded in relevant ways to the physical reality in question: with the kinetic theory of gases, for example, one could see whether such a theoretical gas would obey the empirically known gas laws relating temperature, pressure, and volume. Schrödinger noted that

> [i]f in many various experiments the natural object behaves like the model, one is happy and thinks that the image fits the reality in essential features. If it fails to agree, under novel experiments or with refined measuring techniques, it is not said that one should *not* be happy. For basically this is the means of gradually bringing our picture, i.e. our thinking, closer to the realities.[15]

Schrödinger's own wave-function formalism for a system of particles, which had been fully adopted by all other workers in quantum mechanics, was now generally understood, following the interpretation of German physicist Max Born (1882–1970), as an expression of the probability of finding a given particle at a particular position in space. The wave function, known as the ψ-function, could then be seen as revealing a smeared-out probability field (the relative probabilities were given by the values of ψ^2 rather than of ψ itself). By talking about quantum mechanics in terms of theoretical models, Schrödinger wanted to consider the literal physical status of that particular model: should the wave be understood as a mathematical expression that merely yielded the probability of a particle's being in a specific place? Or should it be understood as representing the particle itself, smeared out over the wave's extent and not "really" in any particular place within it until an observation is made?

This is what he said about the ψ-function:

> That it is an abstract, unintuitive mathematical construct is a scruple that almost always surfaces against new aids to thought and that carries no great message. At all events it is an imagined entity that images the blurring of all variables at every moment just as clearly and faithfully as the classical model does its sharp numerical values. . . . So the latter could be straight-forwardly replaced by the ψ-function, so long as the blurring is confined to atomic scale, not open to direct control. In fact the

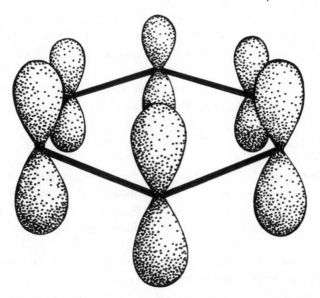

6.6. Quantum chemistry. This diagram depicts a benzene ring with molecular "pi orbitals" at each of its six carbon atoms. These electron "orbitals" are sometimes described by chemists as electron "clouds." They can be regarded as standing-wave patterns or as clouds of probability, following Born: on the latter view, they represent the shape of probability regions within which an electron is most likely to be present. The shape as depicted in such illustrations defines a region within which the electron will be present 90 percent of the time.

> function has provided quite intuitive and convenient ideas, for instance the "cloud of negative electricity" around the [atomic] nucleus, etc.[16]

Schrödinger's talk of "intuitive" concepts refers to picturable, and thence "intelligible," physical notions. These had recently been exploited in the early quantum chemistry of a German-trained American, Linus Pauling (1901–94)—see fig. 6.6.

Schrödinger believed that a "blurred" picture of reality at the atomic scale could be grasped intuitively, but he had doubts about intuition's acceptance of the idea at the scale of everyday objects:

[S]erious misgivings arise if one notices that the uncertainty affects macroscopically tangible and visible things, for which the term "blurring" seems simply wrong. The state of a radioactive nucleus is presumably blurred in such degree and fashion that neither the instant of decay nor the direction, in which the emitted α-particle leaves the nucleus, is well-established. Inside the nucleus, blurring doesn't bother us.[17]

At a larger scale, problems arose.

One can even set up quite ridiculous cases. A cat is penned up in a steel chamber, along with the following diabolical device (which must be secured against direct interference by the cat): in a Geiger counter there is a tiny bit of radioactive substance, *so* small, that *perhaps* in the course of one hour one of the atoms decays, but also, with equal probability, perhaps none; if it happens, the counter tube discharges and through a relay releases a hammer which shatters a small flask of hydrocyanic acid. If one has left this entire system to itself for an hour, one would say that the cat still lives *if* meanwhile no atom has decayed. The first atomic decay would have poisoned it. The ψ-function of the entire system would express this by having in it the living and the dead cat (pardon the expression) mixed or smeared out in equal parts.[18]

This is the first appearance of the famous "Schrödinger's cat paradox." For Schrödinger, though, it was not meant to be a paradox. Instead, it was an example of the unintelligibility, the ridiculousness, of regarding the quantum ψ-function as a kind of blurring of the physical variables that it governed. The role of models in his discussion was to provide a contrast: physical models, Schrödinger held, are always built on the basis of quantities that "are in principle measurable."[19] In the case of quantum blurring, this is not the case, and the counterintuitive cat example shows dramatically why it *should* not be the case. Einstein's assertion that "God does not play dice with the universe" was a less elaborate expression of the same sentiment.

Hence, "reality resists imitation through a model,"[20] Schrödinger decided. Reality transcends whatever a model can grasp, or whatever a model can represent to our understanding. The ψ-function cannot replace a classical model because it does not allow everything to be known or predicted; that was the result of quantum mechanics' focus on probabilities of outcomes rather than on any specific outcome in any particular case. "One notes the great difference over against the classical model theory, where of course from known initial states and with known interaction the individual end states would be exactly known."[21]

But for Schrödinger the chief difficulty lay in the discontinuity that occurred when an observation was made to see what the actual outcome of a process (otherwise known theoretically only through a probabilistic ψ-function) really was: is the cat dead or alive? The ψ-function had been continuous up until that point, but when the observation is made, it collapses to the one actual value among the previous set of possible values. The observer's mere act of perception, then, creates a sharp discontinuity.

Bohr's solution to these difficulties was, according to Schrödinger, to reduce everything to observation and measurement; there was then nothing more to be said. But Schrödinger was not prepared to allow such a minimalist, positivistic interpretation to stand. He proposed that the difficulties might be resolved by attending to the assumption that an observation always occurs at a sharp, specific point in time. "If the formulation could be so carried out that the quantum mechanical predictions did not or did not always pertain to a quite sharply defined point of time, then one would also be freed from requiring this of the measurement results."[22] He went on to remark, "That prediction for sharply defined time is a blunder, is probable also on other grounds,"[23] namely that measurement with clocks is not fundamentally different from any other kind of measurement, which quantum mechanics always regarded as subject to uncertainty.

Schrödinger headed the closing section of his paper "Natural Law or Calculating Device?"[24] This title summed up what for him was the crucial issue: was quantum mechanics truly natural philosophy? Did it tell us how nature is, or did it simply allow us to compute results? Schrödinger evidently saw the second alternative as unsatisfactory, because it would reduce this part of science to nothing more than laboratory instrumentality. Bohr, on the other hand, wanted to make the second alternative a new version of the first: measurement and observation could be at the core of a new kind of natural philosophy. But for Schrödinger, as for Einstein, that just didn't make sense.

VI. Natural Law or Calculating Device?

After the Second World War, quantum mechanics had become firmly established as a standard toolkit for many kinds of physical scientists, including those designing nuclear weapons. Bohr's Copenhagen interpretation was accepted by the vast majority of physicists and chemists, as it has continued to be ever since, largely by default: the issue was irrelevant to users of the practical techniques offered by quantum mechanics, and hence there was little incentive for working scientists to question it. During the 1950s especially, several of the founders of quantum mechanics, most notably Bohr and Heisenberg, continued to lay out the Copenhagen interpretation to general audiences, and the possibility of alternatives was dismissed as fantasy. A useful piece of ammunition provided Bohr's school with apparently irrefutable evidence of the necessity of their arguments: in 1932, the eminent mathematician John von Neumann (1903–57) had produced a proof that a "hidden variables" interpretation of quantum mechanics was an impossibility.

A hidden-variables interpretation would be one in which the results of existing quantum mechanical calculations would remain unaltered, but there would be certain unobservable variables, "hidden" behind the formalism, that could provide a deterministic, rather than merely probabilistic, account of what happened in an experimental situation. The kinetic theory of gases again provides a useful illustration. In that theory, a body of gas is interpreted in terms of its molecules moving through space in all directions and interacting with one another solely through collision. An appropriate statistical treatment of that situation will deal with the expected distribution of velocities among the huge number of molecules in the gas. This approach provides results governing the properties of a gas (such as the relationship between its volume, pressure, and temperature) that can then be compared with experiment. But in principle, and unlike ordinary quantum mechanics, this "classical" theory is based on entirely deterministic processes; the motions and collisions of the molecules are all governed by Newton's laws of motion. If one knew all the mechanical details regarding the positions and motions of every molecule in the gas at a given point in time, one could (again, in principle—this would be a ferocious calculation!) compute the gas's future behavior with perfect accuracy. The only reason that the statistical treatment is favored is that it is practicable, whereas a treatment based on the underlying details of the molecular assemblage is not. A hidden-variables advocate would say that the same might also be true of quantum mechanics; perhaps, if only we could find out what sorts of unknown entities and properties—variables—were analogous to the individual molecules and their behaviors in the kinetic theory of gases, we could develop an interpretation of quantum mechanics that saw its use of statistical methods as a reflection of our ignorance of the precise details of a situation, rather than seeing the indeterminacy as a feature of nature itself.

It soon became clear that von Neumann's proof of the impossibility of a hidden-variables version of quantum mechanics, which would yield the same predictions as the existing theory, was incorrect. The proof was based on several axioms, or foundational assumptions, one of which was suspect. This was already suggested in 1935, and was formally demonstrated in the 1960s. In the meantime, the invocation of the "proof," without much examination, was a common practice of opponents of hidden-variables approaches to quantum mechanics. A circularity in von Neumann's approach was noted by the physicist David Bohm (1917–92) in the 1950s. Bohm published the first significant hidden-variables version of quantum mechanics in 1952, and remained the chief opponent of the Copenhagen orthodoxy until his death.

In *Causality and Chance in Modern Physics* (1957), Bohm referred to "a rather widespread assumption" among physicists such as Heisenberg

> that the human brain is, broadly speaking, able to conceive of only two kinds of things, namely, fields and particles [this is a reference to Bohr's doctrine of complementarity]. The reason generally given for this conclusion is that we can only conceive of what we meet in everyday experience, or at most in experience with things that are in the domain of classical physics where, as is well known, all phenomena fall into one or other of these two classes. Thus, when we come to the microscopic domain, where, as we have seen, neither the field nor the particle concept is adequate, we are supposed to have passed beyond the domain of what we can conceive of.[25]

Bohm buttressed his diagnosis of the situation with a historical parallel. The "mechanistic" physicists of the late nineteenth century had believed that classical physics possessed "an absolute and final validity. . . . What is common to both classical physicists and modern physicists is, therefore, a tendency to assume the absolute and final character of the general features of the most fundamental theory that happens to be available at the time at which they are working."[26]

Bohm's psychological explanation of this attitude towards quantum mechanics ignored the corresponding issue of his own attitude. Much like Einstein, Bohm adhered to an ideal of science that saw it as an enterprise aimed at knowledge of an absolute, "objective" reality that he believed the usual probabilistic quantum mechanics had abandoned. But the instrumental uses to which physical scientists have ever since put the theory have not seemed to demand that kind of knowledge. These days, most serious work on the natural-philosophical underpinnings and implications of quantum mechanics is performed not by physicists but by philosophers of science. Among scientific practitioners themselves, Bohr's attempt to use instrumentality as sufficient grounds for a respectable science has met with a high degree of success.

Conclusion: Making Sense in Science

I. Natural Philosophy and Instrumentality

This book has presented various examples of natural philosophy in the history of science. The hallmark of natural philosophy is its stress on *intelligibility:* it takes natural phenomena and tries to account for them in ways that not only hold together logically, but also rest on ideas and assumptions that seem right, that make sense; ideas that seem natural. Aspects of science that concern such characteristics are sometimes spoken of in the terms of aesthetics, or beauty; the British mathematical physicist Paul Dirac (1901–84) often said that beauty was more important than empirical adequacy in the assessment of a scientific theory: "It is more important to have beauty in one's equations than to have them fit experiment."[1] All of these "contemplative" features—intelligibility, beauty, aesthetics—have long been accorded high cultural status; recall by contrast the low status accorded practical, artisanal endeavors in the Middle Ages.

Natural philosophy, in other words, has always aspired to a lofty cultural place in traditions derived from European models. Furthermore, as the cases discussed in this book have illustrated, intangible, "intuitive" sensibilities have routinely been at the heart of substantive judgments and actions in the history of science; they have always made a difference. It is not the case that scientific theories, world-pictures, and approaches have really been supported and pursued on the basis of instrumental effectiveness alone, with issues of intelligibility having been added afterwards as ornamentation.

To be sure, claims to the intellectual worth of science have often appeared in the form of remarks such as Dirac's on beauty, as well as in the self-presentation of the scientist as the close intimate of a transcendent Nature; in the seventeenth century, Robert Boyle had viewed the natural philosopher as a "priest of nature." Such cultural assertions regarding science tend to play down the role of instrumentality, which is so prominent in other contexts: Niels Bohr wanted to be seen as a profound interpreter of our knowledge of the natural world, rather than just a very good quantum technician, let alone as the man who was smuggled from occupied Denmark to Britain in 1943 because of his expertise in matters relating to the atomic bomb. The intelligibility at the core of natural philosophy has never been inconsequential in the history of the sciences; instead, it has guided and shaped the very content of scientific knowledge, even while that knowledge has relied on appeal to instrumentality as an important complement to science's claims to provide true accounts of nature.

Philosophies of science have commonly tried to address the relationship between the theoretical ideas of science and the concrete experience of the world against which those theories are judged. These have included varieties of "positivism," which in its more extreme forms in the years around 1900 permitted its adherents to deny any real existence to such unobservable, "theoretical" entities as molecular structure or atoms. Kantianism and neo-Kantianism, especially important and well-known in Germany throughout the nineteenth century, was influential on subsequent scientists such as Bohr. Kantianism, the philosophical teachings of Immanuel Kant (1724–1804), held that some of the basic concepts, such as our ideas of three-dimensional space, that people use to make sense of the world are inborn and cannot be dispensed with, even though they might not be literally "true." Neo-Kantianism modified this strong position by suggesting that some of these basic concepts or assumptions might not be necessary and inborn, but could perhaps be chosen freely. A similar viewpoint, known as "conventionalism," was promoted by the mathematician Henri Poincaré (1854–1912). Conventionalism was the position that, while there might not in fact be necessary, unavoidable concepts that we are forced by our natures to use, nonetheless the ways in which we organize our experience of the world depends on using some particular interpretive framework within which to organize it: however, that framework might be replaceable by a quite different one, which would do just as good a job of giving order to the same set of sensory experiences as did the first framework. So Poincaré's view amounted to saying that scientific theories always contain an element of arbitrary, or conventional, interpretive assumptions that cannot be tested experimentally.

Perhaps indirectly encouraged (or else reflected) by such arguments, the natural-philosophical ambitions of science became progressively less evident during the course of the twentieth century. There was an increasing stress on instrumental effectiveness, reflected even in Bohr's observational criteria for the meaningfulness of theories in quantum mechanics, and a pragmatic attitude on the part of working scientists themselves. Piecemeal, instrumental problem solving has become ever more central to the professional identities of many kinds of scientists, and has perhaps weakened the incentive of scientists to present themselves as natural philosophers.

Evidence for this lies in the development of what is sometimes called "technoscience," a term that makes no real distinction between science and technology. Technoscience is problem-oriented, and its most obvious example is that of contemporary work in genomics. Research aimed at discovering the existence of particular genes and determining their precise functions combines with biomedical concerns regarding possible treatments for diseases, yielding an enterprise that allows no real line to be drawn between its natural-philosophical and instrumental components. And in case that example seems unremarkable, here is a complaint against it by a senior biomedical scientist at the University of Michigan. He dislikes the recent efflorescence of life-science institutes at American universities, and calls, in effect, for the purification of such work by eliminating its low-status concern with instrumentality.

> In place of the collective search for evermore powerful explanations, the new view of the life sciences would change the focus to proprietary craft knowledge—knowledge that can be owned and held confidential. This kind of knowledge has no proper role within the academy, an institution centered on the permanent curriculum of arts and sciences and traditionally unconcerned with secrecy—or with capital gains.
>
> The mission of life science institutes emphasizes manipulation and control. Academic sciences, by contrast, are about understanding.[2]

This attitude sheds some light on the way in which the modern ideology of science, which has been a major theme of this book, is maintained. Science is often seen as natural philosophy, and when that conception is foremost, its pursuit for the express purpose of practical utility will appear crass and unintellectual; natural philosophy still carries the higher cultural status. But life science institutes rely on trumpeting the potential benefits of genetic engineering, genomics, and other kinds of techniques that government and industry may have a considerable interest in exploiting: they are usually funded from such sources because the aspect of science that they purvey is that of instrumentality; indeed, the funding of much scientific research throughout the past century has occurred according to the instrumental model. What appears in the article just quoted is a ritual form of purification that maintains the illusion that science is *just* natural philosophy against the apparent anomaly of the new life science institutes.

The other element of the ideology, instrumentality, seldom (if ever) receives a similar treatment because it is of lower status. Instead, the illusion that science is *just* instrumentality receives its routine reinforcement from the way that it is represented in the political arena. Thus, the increased focus placed on scientific education in the United States in the late 1950s was a response to the Soviet launch of the first artificial satellite in 1957; the instrumental capabilities of the Soviet Union translated immediately into measures to improve scientific capacity, because science meant instrumentality. Similarly, the cancellation by the U.S. Congress of the Superconducting Supercollider project in 1993 was justified by its apparent instrumental, or practical, uselessness.

But these two ways of representing science are formally inconsistent with one another: the "science as natural philosophy" perspective sees instrumentality as a direct consequence of the truth of science, while the "science as instrumentality" perspective sees the truth of natural philosophy as being justified by its practical capabilities. As a consequence, and as noted in this book's introduction, while science can be represented in either way, it cannot be represented in both ways simultaneously without a vicious circularity becoming evident. When, in the years after the Second World War, American scientists wanted to encourage government to fund science, they tried to avoid this crucial difficulty by claiming that generous support of "basic science," by which they meant the equivalent of "science as natural philosophy," was the best way to produce practical, especially military, benefits. This was the favored alternative among scientists themselves to an instrumental view of science that would place instrumental achievement ahead of natural philosophy. By using the "natural philosophy" view of science, but putting the stress on instrumental power and suggesting that the latter flowed automatically from natural philosophy, these scientists tried to have things both ways. They exploited the ambiguity in conceptions of what science really was so as to get funding for scientific work that might well have no relation to practical use at all.

But the historical depth of the dual ideology, far from being recent, reaches back to the earliest cases presented in this book. An example from the late sixteenth century makes its longevity clear. That period in Europe saw the beginnings of a new approach to learned knowledge of nature that now made claims both to the explanation of nature and to practical utility, although the exact relationship between the two was open to debate. An English nobleman of the period, Henry Percy, the ninth earl of Northumberland (1564–1632), became known as the "Wizard Earl" because of the natural philosophers who received his patronage. In 1594 he wrote a work in the form of *Advises to His Son,* which included a suggested curriculum of studies. Among the topics was the very practical field of chemistry, which the earl calls "alchemy"—the two were not systematically distinguished at the time. He discusses a particular philosophical understanding of chemical processes, which involves "the method general of all atomical combinations possible in homogeneal substances." He then explains the value of this theoretical approach:

> The application of which doctrine satisfieth the mind in the generation and corruption, as also for the qualities of all substances actually existent . . . which part of philosophy the practice of alchemy doth much further, and in itself is incredibly enlarged, being a mere mechanical broiling trade without this philosophical project.[3]

Or, to paraphrase the earl: in the experimental study of nature, natural philosophy is the part that "satisfieth the mind," and in its absence the work will be "a mere mechanical broiling trade" fit only for workmen. That is, natural philosophy was about supplying intelligible accounts of nature, and the new tools of natural knowledge would be little more than instrumental "trade" if they lacked philosophical content.

In the centuries following the earl of Northumberland's remarks, the role of instrumental efficacy as an element of natural-philosophical intelligibility increased enormously. Earlier chapters have recounted Newton's and Lavoisier's position that if their theories successfully accounted for the phenomena, preferably in a quantitative way, then nothing more should be asked of them. The causes of gravity went beyond the knowable, Newton sometimes implied, while for Lavoisier the underlying nature of chemical principles (whether, for example, they were atomic) could not be learned through currently available investigations. Ultimate philosophical explanations were, in these cases, said to be unattainable; by contrast, instrumental manipulations to produce effects lent their own kind of credibility.

But not all sciences used instrumentality as a means of providing a kind of intelligibility in natural philosophy. Taxonomy in the eighteenth century rested its claims to intelligibility on the ordered layout of its classificatory schemes; Darwin's natural selection seemed to make sense to those who could imagine its operations with the help of Darwin's often-colorful metaphors; Einstein's attitude towards the new quantum mechanics relied on the intuition that only strict causality could yield a universe that would be understandable (Einstein's well-known fondness for thought-experiments as ways of testing or examining theoretical ideas fits the same intuition).

Today, there are still sciences that seem to keep instrumental considerations at arm's length in justifying their assertions. Cosmology, the science of the large-scale structure and behavior of the universe, uses theoretical models that are calibrated by observational, experimentally mediated measurements but make no claims to predictive control over cosmological phenomena—there is really no such thing as experimental, let alone "applied," cosmology. In a different way, evolutionary biology also tends to escape the demands of instrumentality, relying as it does on explanatory structures that are judged primarily by their suitability for organizing complex and often fragmentary evidence (albeit evidence that, again, relies on experimental and observational techniques). Perhaps such sciences are the purest remaining forms of natural philosophy.

II. Teleology and Intelligibility

Other indications of the real historical rootedness of modern science lie in recurrent kinds of arguments. Such arguments display concrete exemplifications of what "intelligible" accounts of nature could, in various times and places, look like. One common way of making sense of nature in the eighteenth century was natural theology (considered above, chapter 4). In the view of many, particularly in Britain, the organization of the natural world, especially but not solely the organic world, could legitimately be used as evidence for an intelligent, designful Creator-God. This God might further be believed to have crafted the universe for the specific benefit of humanity: the world was as it was because God would provide human beings (like lilies of the field) with whatever they needed.

This kind of argument takes for granted the idea that the features and processes found in nature are directed towards goals; things are explained by what they are for. And it is God who made them that way. After Darwin, so the story often runs, such teleological arguments lost all credibility, because natural selection accounted for just the same phenomena without requiring God and conscious design. But modern biologists have sometimes observed that teleological explanation actually remains alive and well in biology, despite Darwin. The only difference is that God is not invoked; natural selection plays the same role, albeit largely implicitly. Teleological arguments in biology remain acceptable in practice as long as it is understood that they could always be reexpressed, usually in a rather unwieldy fashion, in the terms of natural selection. This is testimony to Darwin's success in taking the issues and themes presented to him by natural theology and translating them wholesale into the different idiom of natural selection, but it can also be seen as a mark of the effectiveness of the original natural-theological arguments.

A similar translation of teleological intelligibility can be found in the case of a theory, or theoretical approach, that originated in modern cosmology. The "anthropic cosmological principle" was first presented in its received form in 1974 by a professional cosmologist and has received much discussion since, often in popularized form. Its major statement came in 1986, in a book by John Barrow and Frank Tipler. The central idea is to include the existence of ourselves, human beings, as a part of our understanding of the way the universe is. Human existence is seen not simply as an accidental fact about the universe, but as a somehow *necessary* fact—it must be the case that human beings can exist in the kind of universe we inhabit, because if it were not, we wouldn't be here to discuss it. It can therefore be inferred, the argument goes, that the universe has properties that are suited to our existence; the possibility of the natural emergence of human beings is a necessary condition of the properties of our universe.

The anthropic cosmological principle, therefore, turns the position of natural theology on its head in a way comparable to the effect of Darwin's natural selection. But it is of much greater scope, because it applies not only to our understanding of organic life, but to the properties of the inorganic world as well. Barrow and Tipler write:

> At first sight, such an observation [i.e., that the universe needs to have made human intelligence possible] might appear true but trivial. However, it has far-reaching implications for physics. It is a restatement, of the fact that any observed properties of the Universe that may initially appear astonishingly improbable, can only be seen in their true perspective after we have accounted for the fact that certain properties of the Universe are necessary prerequisites for the evolution and existence of any observers at all. The measured values of many cosmological and physical quantities that define our Universe are circumscribed by the necessity that we observe from a site where conditions are appropriate for the occurrence of biological evolution and at a cosmic epoch exceeding the astrophysical and biological timescales required for the development of life-supporting environments and biochemistry.[4]

Other properties of the universe that can be related to the existence of humanity include certain features of our solar system (cf. Newton's theistic view, above, chapter 1), the chemical properties of carbon atoms (to produce life), and even the values of basic physical constants like the mass ratio of protons and electrons (which enable those chemical properties). There are many such examples, and a number of different versions of the anthropic principle, all of which can be expressed in terms of teleological reasoning: the universe is the way it is because it needs to produce us. But in this argument, God is not required.

Nonetheless, there is a very similar general argument in the annals of the earth-sciences that does include God. In the late eighteenth century, the Scot James Hutton (1726–97) produced a *Theory of the Earth* that attempted to incorporate geological, paleontological, and astronomical evidence. His idea was to present an account of the earth as a system, a coherent, self-sustaining "machine" the various features of which cooperated with one another to create a stable environment. Hutton's earth was not one that developed over time; instead, it existed in a fundamentally unchanging balance of processes whereby the overall appearance and characteristics of the earth were always the same, despite changing details such as the precise shapes and distribution of the continents and oceans. He said that there was no geological evidence of the earth ever having been essentially different from the way it is now, and that there were no known long-term processes that enabled prediction of essential change in the future: "we find no vestige of a beginning,—no prospect of an end."[5]

What led him to this ideal image, apart from his reading of the always-interpretable evidence, was a teleological premise remarkably similar to that of the anthropic principle:

When we trace the parts of which this terrestrial system is composed, and when we view the general connection of those several parts, the whole presents a machine of a peculiar construction by which it is adapted to a certain end. We perceive a fabric, erected in wisdom, to obtain a purpose worthy of the power that is apparent in the production of it.[6]

So Hutton claimed that the earth is obviously the work of a wise Creator. Furthermore, he specified the crucial feature of its design: "This globe of the earth is a habitable world; and on its fitness for this purpose, our sense of wisdom in its formation must depend."[7] This was Hutton's own "anthropic principle," and he used it in his book to make a host of inferences regarding those properties and

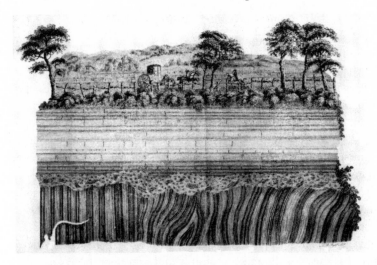

7.1. Hutton conceived of the earth as a system whose past was recorded in its rock strata, and that could be understood by combining that evidence with the principle that God had made the earth for human habitation. Here, in Hutton's Theory of the Earth *(1795), we see a cross-section of the earth's surface displaying the layered rock, while above everything else are human beings, literally the pinnacle of creation.*

processes of the earth that flowed from it. His central thesis of the earth's stability, taken as a system, rested on the necessity of such a world to sustain human life without periodically destroying it (see fig. 7.1). Hutton's ideas amounted to a program for actually doing geological research, and in his discussions of such matters as coal strata his natural philosophy was juxtaposed directly with things of great utilitarian interest. Some decades later, in the quite similar approach of Hutton's fellow Scot Charles Lyell, Hutton's "uniformitarianism" was to become enormously consequential in the development of modern geology. As a way of making sense of the world, teleology has always been attractive, and the expulsion of God from the usual forms of scientific explanation has really only changed the idiom. Nature, for whatever reason, often seems to make sense on the assumption that it has purpose.

III. Proof and Understanding in Mathematics

Another form of intelligibility involves accepting a belief because it seems to follow clearly and inevitably from other accepted beliefs. This is the model of ancient Greek mathematics as found in Euclid's *Elements*. Once all of Euclid's starting principles are accepted, the truth of the theorems deduced from them must also be accepted because, on this view, the logical steps leading from first principles to conclusions force assent: no rational person can deny the conclusions once the steps leading to them have been laid out. The deductive conception of scientific argument in general, not restricted to mathematics, is older even than Euclid (c. 300 B.C.), going back at least to Aristotle. But the mathematical version makes clear what has always been at stake in this ideal of knowledge: the formal deductive proof is a way of showing not simply that some assertion *is*, as a matter of fact, true, but that it *must* be true. For instance, Euclid provides a demonstration of the proposition that "[i]n isosceles triangles the angles at the base are equal to one another, and, if the equal straight lines be produced further, the angles under the base will be equal to one another."[8] An isosceles triangle is a triangle two of whose sides are of equal length, with the third understood to be its base; anyone who knows this might conclude, perhaps from an intuition of symmetry, that the proposition is trivially true (fig. 7.2). Certainly, Euclid could have anticipated that few would doubt its truth. And yet he bothered to demonstrate that the conclusion must follow from the basic principles that he had laid down earlier in the *Elements*. The real point of a Euclidean demonstration was to show *why* a true statement was true.

The recurrent popularity of this model of knowledge seems to stem (judging from the frequent paeans of praise that it has received throughout Western intellectual history) from the apparent certainty of such proofs—but not only from their certainty. A mathematician didn't just know as a matter of certain fact that something happened to be true; he also understood its truth. In other words, such a proof made a mathematical truth intelligible.

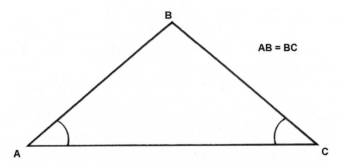

7.2. *The two base angles of an isosceles triangle, according to Euclid's treatment, are necessarily equal to one another. In the dominant mathematical tradition stemming from such Greek authors, this statement of equality is not a mere matter of fact, but a demonstrable truth about how things* must *be, given certain prior assumptions.*

There have always been competitors to this mathematical model of scientific truth, however. One of the more persistent ones is called "hypothetico-deductive." In the latter, a conjectural explanation for some phenomenon is proposed as a hypothesis; for instance, the kinetic theory of gases, which posits the existence of molecules in rapid motion as the reality lying behind the manifest properties of bodies of gas (see p. 171, above). The hypothesis implies certain consequences (such as the relationship between pressure, temperature, and volume for a gas composed of such molecules), and those consequences can then be compared with the experimentally determined properties of real gases. As such seventeenth-century natural philosophers as Descartes and Boyle knew perfectly well, a close correspondence between predictions deduced from a hypothesis and actual empirical results in no way proved the truth of the hypothesis; perhaps some other hypothesis might have done equally well. But some version of this hypothetico-deductive approach to explanation has frequently been used ever since as a way of building theories that can at least be tested, even if they can never be proved with certainty.

Einstein regarded the hypothetico-deductive approach to physics as decidedly inferior. He called the ideas that resulted from it "constructive theories," because they were theories built on the foundation of speculative and unprovable assumptions. Einstein's preferred alternative was the "theory of principle," which was derived from starting principles that were justified not in retrospect, once experiments had confirmed their consequences, but at the outset, on the basis of their correspondence to general experience. Classical thermodynamics was a prize example of a theory of principle, because it was widely regarded as being rooted in easily accepted principles such as "heat flows spontaneously only from a hotter to a colder body." In 1905, Einstein had presented his theory of special relativity as a theory of principle, including the principle of relativity itself (based on much negative experimental knowledge regarding electromagnetic as well as mechanical phenomena), and the similar principle of the invariability of the speed of light in free space. The high status that Einstein accorded to theories of principle evidently flowed from their firmer integration into an overall conception of how the world works; the principles he had in mind were ones that would be consistent with, and a part of, already-existing and quite fundamental knowledge about the physical universe. Such principles were therefore, he thought, practically guaranteed to reveal further truths hidden in the entangled web of reality.

Einstein's ultimate model was, of course, the geometrical theorem rooted in supposedly self-evident first principles. Like Aristotle, he adhered to a conception of scientific demonstration that attempted to make the natural sciences as much like mathematical deduction as possible. But even in pure mathematics itself, the transparency and intelligibility of deductive demonstrations is sometimes less than obvious. Sextus Empiricus, a Greek philosopher writing in the second century, long ago suggested potential problems with the deductive proofs of mathematics. His arguments were, in pragmatic terms, rather far-fetched. He said that a geometrical theorem, demonstrated deductively from unquestioned first principles, might not be able to show the necessary truth of its conclusion; he noted that when a mathematician works through a proof to check its validity, mistakes are always possible, mathematicians being fallible human beings. Even after many repetitions of this careful step-by-step examination, errors can never be absolutely ruled out. Therefore, Sextus concluded, there is no completely certain knowledge, even in mathematics.

The level of skepticism involved in that argument was seldom taken seriously by anybody. But even the most ancient philosophy can sometimes come back with a vengeance. In 1993 a Princeton mathematician, Andrew Wiles, announced that he had developed a proof of a famous theorem first enunciated in the seventeenth century by the Frenchman Pierre de Fermat (1601–65). Long known as "Fermat's last theorem," it was a claim about number theory that Fermat scribbled in a margin without giving any proof for it. Analogous to the famous theorem of Pythagoras concerning the squares on the sides of right-angled triangles, Fermat's proposed theorem asserts that the expression $x^n + y^n = z^n$ has no solutions that are positive whole numbers if n is greater than two (when $n = 2$, the equation expresses the relation among the sides of a right triangle according to the Pythagorean theorem).

Leaving aside the details, what is striking about Wiles' solution to the problem—his proof of Fermat's theorem—is that, more than three centuries after Fermat's death, this celebrated proof ran to more than one hundred pages when published, and relied on the detailed, highly specialized work of other mathematicians to underpin several of its component parts. And as a result, much as Sextus Empiricus would have expected, no one mathematician was in a position to confirm its validity as beyond doubt. When the official referees for the mathematical journal in which Wiles wanted to publish his proof worked their way through those parts of it that they were most competent to assess, they asked Wiles for clarification on various points. One of those points, concerning chapter 3 of Wiles's paper, turned out to be a serious difficulty. Wiles commented on his reaction:

> I couldn't immediately resolve this one very innocent looking question. For a little while it seemed to be of the same order as the other problems, but then sometime in September [1993] I began to realise that this wasn't

just a minor difficulty but a fundamental flaw. . . . Even explaining it to a mathematician would require the mathematician to spend two or three months studying that part of the manuscript in great detail.[9]

Wiles eventually found a way around the problem as a result of collaboration with another mathematician, and in late October 1994 the paper, with appropriate changes, was finally ready to be published. The rest of the professional mathematical community was now prepared to be satisfied of the proof's validity—even though the practical difficulties in verifying it, unaided, on the basis of simply reading through the paper were evidently insuperable. "Proof" was here an intellectual accomplishment that required a social achievement: relevant people had to agree that a complex argument was adequate to its job of absolute proof.

Some years before, however, the social operation of the mathematical community had already revealed itself in open controversy over a proof's validity. This was the celebrated "four-color map" theorem, originally proposed in 1852. It considers any plane, two-dimensional map, with its surface divided up by lines continuous with one another (no unconnected "islands"), into any number of distinct areas of any shapes whatsoever (e.g. fig. 7.3). The theorem holds that such a map can always be painted with no more than four different colors, such that no directly adjacent areas are of the same color.

There had been many attempts to provide a formal proof for this theorem. But the most notable came in the 1970s, and it owed its celebrity to its use of computer assistance. The mathematicians who announced their result in 1976, Wolfgang Haken, Kenneth Appel, and John Koch, had developed an elaborate technique to prove the theorem. It involved reconceptualizing the structure of these kinds of plane maps in terms of configurations of elements linked to one another, analogous to a web of electrically charged nodes. What they wanted to do was to show that if you postulate the existence of a map with the smallest number of distinct areas possible in which *five* colors are required, you will inevitably find that such a thing will be reducible (reconfigurable) into a simpler but equivalent map in

7.3. *All the enclosed areas on this kind of "map" can be shaded with four or fewer colors such that no two areas of the same color border one another. But proving it as a general theorem is not easy.*

which there is a smaller number of areas, but that still requires no fewer than five colors. Such a result would amount to a self-contradiction, since the first map had already been defined as one with the smallest number of areas necessary to require five colors. This kind of proof, with plenty of examples in ancient Greek mathematics, is usually described by the Latin phrase *reductio ad absurdum:* show that if you assume the truth of some statement, you will be led to a self-contradiction (the absurdity); you can then conclude that the original assumption must not be true after all, and is therefore false. In the present case, it would be shown that no map requires five colors, and since it was already known from earlier work that six and above were never required, and that fewer than four were insufficient, clearly four was the maximum number of colors that would ever be needed.

In 1976, the mathematicians determined, through human endeavor, a certain subset of all the possible configurations of five-color maps. This subset consisted of all the configurations that were not

obviously and trivially reducible in the manner just described. Their
task was then to show that every single one of these 1,936 nontrivial
cases could in fact be reduced in just the same way. The amount of
work required exceeded practical human capacity, and they used a
computer to slog through every one of the configurations for them.
The outcome was that the computer program confirmed reducibil-
ity in every instance. Accordingly, the proof had been accomplished.

But had it? If the purpose of a mathematical proof is to show *why*
something must be the case—to enable the mathematician to understand its necessity—had this piece of work accomplished the
task? Many mathematicians at the time, and subsequently, believed
that Haken et al.'s "proof" was no such thing, and should not count
as a mathematical proof at all. One mathematician wrote of the result that it

> doesn't give a satisfactory explanation *why* the theorem is true. This is
> partly because the proof is so long that it is hard to grasp (including the
> computer calculations, impossible!), but mostly because it is so appar-
> ently structureless. The answer appears as a kind of monstrous coinci-
> dence.[10]

A philosopher of mathematics elaborated on this viewpoint by referring to a property of "surveyability," by which he meant the capacity of a formal proof to be grasped in its entirety by the human
mind.[11] Another mathematician described the advent of computer-
assisted proofs as "dangerous," and said, regarding the Haken four-
color proof, that he would be "much happier if there were a written
proof that I could comprehend in my usual fashion."[12] But lack of
surveyability can apply to any very lengthy and elaborate proof,
whether or not conducted using computers, as Sextus Empiricus
knew: there is a report that, following a presentation at Berkeley of
the proof of the four-color map theorem, "The people over forty
could not be convinced that a proof by computer could be correct,
and the people under forty could not be convinced that a proof that
took 700 pages of hand calculations could be correct."[13]

IV. Natural Philosophy and Instrumentality Revisited

Twentieth-century philosophers of science sometimes saw themselves distinguished into two main camps: on the one hand, "realists" reckoned that the instrumental successes of science indicated that the hidden entities scientific theories talked about (genes, protons, superstrings) most likely really existed. "Instrumentalists," on the other hand, held that there was no compelling reason to make that inference at all—science "works," and that's all that can be said; if the associated theories referred to things called "protons," that shouldn't be taken as anything other than a manner of speaking that enabled the calculation of accurate predictions. This book, by contrast to either of those positions, shows science as a practical human enterprise that has grown up over the past few centuries by developing an intimate relationship between, not two opposed philosophies, but two distinct practical endeavors: natural philosophy, the goal of which is to make sense of the world, and instrumentality, which aims at creating means of material control. The history of science is in large part the story of how those two, while never quite acknowledging the fact, have interwoven and accommodated themselves to each other.

There is nothing unusual in representing science as a system made up of two distinguishable endeavors. Paired elements of science that have long been identified in philosophical discussions, as well as by scientists themselves, are theory and experiment. The first is intellectual and suitable for "theoreticians," while the second is practical, technical and hard-nosed, the province of "experimentalists." The most common way of describing their relationship has been to give theory priority, with experiment playing the role of confirming or falsifying the predictions of theory. In the past few decades, various changes have been rung on such dichotomies: the philosopher Stephen Toulmin, in *Foresight and Understanding* (1961), described "foresight," or "forecasting," as "a craft or technology, an application of science rather than the kernel of science itself." For Toulmin, true science was about understanding *why* forecasts worked;

successful forecasts, after all, can often be made with little grasp of the reasons for their success. Another philosopher of science, Ian Hacking, wrote the influential *Representing and Intervening* (1983), in which the connection between the intellectual content of science, taken to be the intellectual "representing" of the world, is coupled with experimental "intervening," in which the world is materially acted upon, thereby bearing witness to the reality of the associated representations. At the same time, some scholars have begun to focus more on scientific practice and less on scientific theories; Peter Galison, for example, has presented modern physics as an enterprise constituted from distinct traditions of practice, among them a theoretical and an experimental tradition, which continually negotiate, or "trade," with one another. The many variants of the theory/experiment pairing in ideas about the structure of science all have one thing in common, however: each is understood to contribute to an endeavor of which the essential aim is natural-philosophical rather than instrumental.

As we have seen in earlier chapters, the competing demands of these two interwoven dimensions of modern science tend to talk past one another. The overriding and characteristic feature of natural philosophy, intelligibility, has routinely been used as a double of the chief feature of instrumentality, namely efficacy. Sometimes intelligibility is granted intellectual priority over efficacy, while sometimes efficacy is given material priority over intelligibility: the former is perhaps the case in fields such as modern cosmology or evolutionary biology, while the latter seems to be the case in everyday uses of quantum mechanical techniques by physicists and chemists.

A nice illustration of these distinct values can be found in a non-scientific endeavor with clearly delineated goals: the game of chess. In recent decades, computer programs have been developed with ever greater practical abilities in chess, but the greatest achievements have resulted from using "brute force" methods to analyze endgame positions. Chess endgames are situations where few pieces are left on the board. Powerful computers have in many cases been able to produce exhaustive analyses of endgames, sometimes showing certain endgame positions to be drawn, or forced wins for one side, that the most skilled human chess players had evaluated differently. Some of these computer-analyzed determinations, reminiscent of the mysteries of the four-color map problem, have been dubbed "God endgames." Although the computer (when trusted) is said to have shown through exhaustive analysis that, with correct play, one side must win, the technique that achieves the quickest forced win is reckoned by the strongest human chess players to be literally *unintelligible*. No grandmaster could explain why a particular move should be the best in some particular position, even though the computer analysis decides that it is (and the advantage of chess being a game is that any doubters regarding the effectiveness of the computer's moves could attempt to beat it in play . . .). A God endgame is thus an endgame that only God could understand; it is simply not intelligible to human beings.

Such cases are analogous to the situation in mathematics: as we saw, the truth of a theorem is a different matter from a *proof* of that theorem; in chess, the win in an endgame is different from understanding how to do it. Understanding is an important issue in the minor art of chess endgame problems, in which people compose endgame positions where the correct moves illustrate a winning theme or idea. In science, efficacy relates to intelligibility in an analogous way. We say that science works, and we also say that it helps us to understand the world. But just as a computer chess analysis sometimes cannot provide an understanding of chess efficacy even to a grandmaster, so a successful scientific technique need not yield an intelligible account of nature. However, practical scientific success without associated natural-philosophical intelligibility might well not seem like science at all, but just an ill-understood accomplishment.

Ironically, views of science and scientists in modern culture sometimes regard widespread failure to comprehend an idea or a theory as a sign of its profundity. In the early decades of the twentieth century, it was popularly held that only a very few people in the world could understand Einstein's 1915 general theory of relativity. On this view, only enormously clever people were capable of it, and this therefore showed what a brilliant theory it was. Of course, the (apocryphal) story is a far cry from saying that the theory was unintelligible; it was just hard. If leading physicists had alleged that it failed to make any sense, they would have been denying to it a central requirement of a genuine scientific theory—that it possess natural-philosophical content.

Almost the opposite point is shown by a much more recent theory, known as "string theory." This elaborate and complex area of mathematical physics, with its ten spatial dimensions, attempts to unify relativity theory and quantum field theory, and has become important despite the notorious fact that it is said to admit, as yet, of no known experimental test. Here, doubts attaching to its proper status as a scientific theory are not due to a claim that it is unintelligible, but to what is in effect its apparent distance from any conceivable experimental instrumentality. So both elements, natural philosophy *and* instrumentality, seem to be needed to make a proper scientific theory.

But science's natural philosophy, as this book has illustrated, requires intelligibility (perceived or reputed) to render it plausible. The natural-philosophical assertions made by science are not based simply on scientists' "proofs" about the way the world is. They are judged from the start on whether they make sense, and the controversies over that very issue, revealed again and again in the history of science, show how "making sense" depends on who is doing the judging, and in what cultural circumstances. The world pictures that we believe in owe much more to what we find plausible than to the way the world "really" is: their acceptance, rather than being determined by the natural world itself, depends on the ways in which we choose to live in the world.

A further point in conclusion: throughout this book, I have spoken of "science" and the ways in which it has shown, over the past four centuries or so, particular cultural and intellectual features. Science appears in this light as a rich and diverse cultural enterprise, but as such it is in origin a specifically "Western" or European one. Non-Western cultures have long had their own conceptions of the natural world, as well as developing countless techniques for inter-acting with and making use of it. It is a peculiarity of the specific Eu-ropean social institution that we call science that it has, in various historically contingent ways, combined those two things into what it presents as a single endeavor—but which is more correctly de-scribed as an ideology. When European expansion, especially in the nineteenth and twentieth centuries, exported "science" to other parts of the world, its reception in other settings often displayed the very cracks and incoherences that the ideology of science continues to suppress.

Notes

Chapter One

1. René Descartes, *The World and Other Writings*, trans. Stephen Gaukroger (Cambridge: Cambridge University Press, 1998), p. 26.

2. Galileo Galilei, *Dialogue Concerning the Two Chief World Systems*, trans. Stillman Drake (Berkeley: University of California Press, 1967), p. 234.

3. Thomas Hobbes, *Leviathan* (London: Dent Dutton, Everyman's Library, 1914 [1651]), p. 371; Hobbes, *Leviathan, with Selected Variants from the Latin Edition of 1668*, ed. Edwin Curley (Indianapolis: Hackett, 1994 [1651]), p. 462.

4. Francis Bacon, *The New Organon*, ed. and trans. Lisa Jardine and Michael Silverthorne (Cambridge: Cambridge University Press, 2000), bk. 1, aph. 34.

5. Ibid., aph. 33.

6. Descartes, *The World*, p. 21.

7. Ibid.

8. René Descartes, *The Philosophical Writings of Descartes*, trans. Robert Stoothoff, Dugald Murdoch, and John Cottingham, 3 vols. (Cambridge: Cambridge University Press, 1985–91), vol. 1, p. 256.

9. Ibid.

10. Ibid., p. 132.

11. Descartes, *The World*, p. 24.

12. Christiaan Huygens, *Oeuvres complètes de Christiaan Huygens*, 22 vols. (The Hague: Nijhoff, 1888–1950), vol. 19, p. 631. All unattributed translations in this book are mine.

13. Both quoted in Alexandre Koyré, *From the Closed World to the Infinite Universe* (Baltimore: Johns Hopkins University Press, 1957), pp. 178–79.

14. Descartes, *Philosophical Writings*, vol. 3, p. 372.

15. Isaac Newton, *Opticks, or A Treatise of the Reflections, Refractions, Inflections &
Colours of Light* (New York: Dover, 1952), p. 398. This edition reproduces the text
of the fourth edition, published in 1730, but the quoted material dates from the
edition of 1717/18, having first appeared in a Latin version in 1706.

16. Ibid., p. 399.

17. Ibid., p. 401.

18. Ibid.

19. Ibid., pp. 401–2.

Chapter Two

1. Isaac Newton, *Opticks, or A Treatise of the Reflections, Refractions, Inflections &
Colours of Light* (New York: Dover, 1952), p. 381.

2. *Histoire de l'Académie Royale des Sciences, année MDCCXVIII* (Paris, 1741), p. 36;
partially trans. in Alistair Duncan, *Laws and Order in Eighteenth-Century Chemistry*
(Oxford: Clarendon Press, 1996), p. 157.

3. *Histoire*, p. 37, my trans.; I vary slightly from Duncan, *Laws and Order*, p. 157.

4. Georges-Louis Leclerc, comte de Buffon, *Histoire naturelle, générale et par-
ticulière*, vol. 1 (Paris, 1749), "Premier discours," p. 52, my translation; cf. other
English versions of this passage in John Lyon and Phillip R. Sloan, eds., *From
Natural History to the History of Nature: Readings from Buffon and His Critics* (Notre
Dame: University of Notre Dame Press, 1981), p. 122, and Jacques Roger, *Buffon: A Life in Natural History*, trans. Sarah Lucille Bonnefoi (Ithaca: Cornell University Press, 1997), p. 85.

5. "Colonizing the mind" is a striking figure borrowed from Mary Douglas,
Implicit Meanings: Essays in Anthropology (London: Routledge & Kegan Paul, 1975),
p. xx.

6. "Extrait des Registres de la Société Royale de Médecine, du 3 Juillet
1789," in A.-L. de Jussieu, *Genera plantarum*, with an introduction by Frans A.
Stafleu (Weinheim: J. Cramer, 1964), pp. 12–24, on p. 12.

7. Ibid.; cf. Jussieu, *Genera plantarum*, "Introductio," p. xxxv, where Jussieu
himself speaks similarly about his "natural method."

8. Quoted in Stafleu, introduction to Jussieu, *Genera plantarum*, p. xx, my
translation.

9. Quoted in Simon Schaffer, "Herschel in Bedlam: Natural History and

Stellar Astronomy," *British Journal for the History of Science* 13 (1981), pp. 211–39, on p. 217.

10. Trans. in Martin J. S. Rudwick, *Georges Cuvier, Fossil Bones, and Geological Catastrophes: New Translations and Interpretations of the Primary Texts* (Chicago: University of Chicago Press, 1997), p. 19.

11. Ibid.

12. Ibid., p. 19 n. 9.

13. Ibid., p. 22.

Chapter Three

1. Georg Ernst Stahl, *Philosophical Principles of Universal Chemistry* (London, 1730), p. 4.

2. Quoted in Charles Coulston Gillispie, "The *Encyclopédie* and the Jacobin Philosophy of Science: A Study in Ideas and Consequences," in *Critical Problems in the History of Science*, ed. Marshall Clagett (Madison: University of Wisconsin Press, 1959), pp. 255–89, original quote on pp. 258–59 (my trans.), from *Encyclopédie*, vol. 3, p. 415.

3. Antoine-Laurent Lavoisier, *Elements of Chemistry*, trans. Robert Kerr (Edinburgh, 1790; facs. reprint New York: Dover, 1965), pp. xv–xvi. Cf. the opening sentence of Jane Austen's *Pride and Prejudice* (1813).

4. Lavoisier, *Elements*, p. xvi.

5. Ibid., p. xviii.

6. Ibid., pp. xx–xxi.

7. Ibid., p. xxxvii.

8. See ibid., p. 218.

9. Ibid., p. 175.

10. Quoted in Simon Schaffer, "Priestley's Questions: An Historiographic Survey," *History of Science* 22 (1984), pp. 151–83, on p. 164.

11. Quoted in Jan Golinski, " 'The Nicety of Experiment': Precision of Measurement and Precision of Reasoning in Late Eighteenth-Century Chemistry," in *The Values of Precision*, ed. M. Norton Wise (Princeton: Princeton University Press, 1995), pp. 72–91, on p. 81, from a contemporary English trans.

12. Quoted in ibid., p. 84.

13. Joseph Priestley, *Experiments and Observations on Different Kinds of Air*, 3 vols. (London, 1775), vol. 2, sec. 3, extracted in Henry M. Leicester and Herbert S.

Klickstein, *A Source Book in Chemistry, 1400–1900* (New York: McGraw-Hill, 1952), p. 113.

14. Ibid.

15. Ibid, pp. 116, 117, 117, 118, 119.

16. Ibid., pp. 119–20.

17. Ibid. p. 116.

18. Quoted in Arnold Thackray, *John Dalton: Critical Assessments of His Life and Science* (Cambridge, MA: Harvard University Press, 1972), p. 72.

19. Ibid., p. 71.

20. Ibid., p. 73.

21. Lavoisier, *Elements*, p. xxiv.

22. Quoted in Thackray, *John Dalton*, p. 85.

Chapter Four

1. Charles Darwin, *The Correspondence of Charles Darwin*, ed. Frederick Burkhardt et al. (Cambridge: Cambridge University Press, 1985–), vol. 7, p. 423.

2. William Paley, *Natural Theology; or, Evidences of the Existence and Attributes of the Deity* (London, 1809), p. 441.

3. Charles Darwin, *The Autobiography of Charles Darwin, 1809–1882*, with original omissions restored, ed. Nora Barlow (London: Collins, 1958), p. 59.

4. Charles Darwin, *On the Origin of Species*, a facsimile of the 1st ed., ed. Ernst Mayr (Cambridge, MA: Harvard University Press, 1964), p. 287.

5. Ibid.

6. Anonymous reviewer quoted in Darwin, *Correspondence*, vol. 8, p. 10 n. 16.

7. Quoted in ibid., p. 63 n. 16.

8. Quoted in Crosbie Smith and M. Norton Wise, *Energy and Empire: A Biographical Study of Lord Kelvin* (Cambridge: Cambridge University Press, 1989), p. 597.

9. Darwin, *Origin*, ed. Mayr, p. 189.

10. Ibid., p. 188; emphasis added.

11. Ibid., p. 117.

12. Ibid., p. 191.

13. Ibid.

14. Reported by Darwin to Lyell, 10 Dec. 1859, in Darwin, *Correspondence*, vol. 7, p. 422.

15. Darwin, *Origin*, ed. Mayr, p. 184.

16. Richard Owen, "Darwin on the Origin Of Species" (1860), reprinted in David L. Hull, *Darwin and His Critics: The Reception of Darwin's Theory of Evolution by the Scientific Community* (Chicago: University of Chicago Press, 1973), pp. 175–213, quote on p. 198.

17. Adam Sedgwick, "Objections to Mr. Darwin's Theory of the Origin of Species" (1860), reprinted in Hull, *Darwin and His Critics,* pp. 159–66, quote on p. 165.

18. See Darwin to Lyell, 25 Nov. 1859: "I will certainly leave out Whale and Bear," in Darwin, *Correspondence,* vol. 7, p. 400; also Darwin to W. H. Harvey, 20–24 Sept. 1860, "as it offended persons I struck it out in 2d Edition," in ibid., vol. 8, p. 371. Owen criticized Darwin on the question in person, as reported in Darwin to Lyell, 10 Dec. 1859, ibid., vol. 7, pp. 422–23.

19. Charles Darwin, *The Origin of Species,* based on 2nd ed., ed. Gillian Beer (Oxford: Oxford University Press, 1996), p. 150.

20. W. H. Harvey to Darwin, 24 Aug. 1860, in Darwin, *Correspondence,* vol. 8, p. 323.

21. See J. D. Hooker to Darwin, 2 July 1860, in ibid., pp. 270–72 and p. 271 n. 9.

22. Thomas Vernon Wollaston, review of *Origin* (1860), reprinted in Hull, *Darwin and His Critics,* pp. 127–40, quote on p. 139 (emphasis in original).

23. Charles Coulston Gillispie, *The Edge of Objectivity: An Essay in the History of Scientific Ideas* (Princeton: Princeton University Press, 1960), p. 304.

24. Samuel Haughton, "Biogenesis" (1860), reprinted in Hull, *Darwin and His Critics,* pp. 217–27, quote on p. 223.

25. Ibid.

26. William Benjamin Carpenter, "Darwin on the Origin of Species" (1860), reprinted in Hull, *Darwin and His Critics,* pp. 88–114, quote on p. 109.

27. Joseph Dalton Hooker, review of *Origin* (1859), reprinted in Hull, *Darwin and His Critics,* pp. 81–85, quote on p. 85.

28. H. G. Bronn, review of *Origin* (1860), reprinted in Hull, *Darwin and His Critics,* pp. 120–24, quote on p. 124.

29. Frederick Wollaston Hutton, "Review of the Origin of Species" (1860), partially reprinted in Hull, *Darwin and His Critics,* pp. 292–300, quote on p. 299.

30. Darwin to W. H. Harvey, 20–24 Sept. 1860, in Darwin, *Correspondence,* vol. 8, p. 371.

31. Quoted in Robert M. Young, "Darwin's Metaphor: Does Nature Select?" in Young, *Darwin's Metaphor: Nature's Place in Victorian Culture* (Cambridge: Cambridge University Press, 1985), pp. 79–125, on p. 96.

32. Sedgwick to Darwin, 24 Nov. 1859, in Darwin, *Correspondence*, vol. 7, p. 397.

33. Darwin, *Origin*, ed. Mayr, p. 482.

34. Ibid., p. 490.

Chapter Five

1. Pierre Duhem, *The Aim and Structure of Physical Theory*, trans. Philip P. Wiener from French 2nd ed., 1914 (Princeton: Princeton University Press, 1954), pp. 70–71.

2. E.g., in Michael Faraday, *Experimental Researches in Electricity*, 3 vols. (London, 1839–55), vol. 3, para. 3258 (material originally published 1852); see also other passages quoted in L. Pearce Williams, *Michael Faraday: A Biography* (New York: Basic Books, 1965), pp. 450–54.

3. Faraday, *Experimental Researches*, vol. 3, para. 3306, with a footnote quoting from Newton's third letter to Bentley.

4. Quoted in Crosbie Smith and M. Norton Wise, *Energy and Empire: A Biographical Study of Lord Kelvin* (Cambridge: Cambridge University Press, 1989), p. 123.

5. Ibid.

6. Quoted in ibid., p. 234 (my emphasis).

7. Faraday, *Experimental Researches*, vol. 3, para. 2166.

8. Quoted in Smith and Wise, *Energy and Empire*, p. 258.

9. Faraday, *Experimental Researches*, vol. 3, para. 3306.

10. Quoted in John Hendry, *James Clerk Maxwell and the Theory of the Electromagnetic Field* (Boston: Hilger, 1986), p. 163.

11. Thomson to Helmholtz, in Silvanus P. Thompson, *The Life of William Thomson, Baron Kelvin of Largs*, 2 vols. (London, 1910), vol. 1, pp. 514–15; cf. Smith and Wise, *Energy and Empire*, p. 418.

12. Quoted in Smith and Wise, *Energy and Empire*, p. 423.

13. William Thomson and Peter Guthrie Tait, *Treatise on Natural Philosophy* (Oxford: Clarendon Press, 1867), vol. 1, p. v.

14. James Clerk Maxwell, "Attraction," *Encyclopaedia Britannica*, 9th ed., 25 vols. (New York: Scribners, 1878–89), vol. 3, pp. 63–65, on p. 64.

15. Ibid.

16. James Clerk Maxwell, *Scientific Papers*, 2 vols., ed. W. D. Niven (Cambridge: Cambridge University Press, 1890), vol. 1, p. 452.

17. Ibid., p. 468.

18. Ibid., p. 486.

19. Ibid., p. 528.

20. Ibid., p. 538.

21. Ibid., p. 564.

22. Quoted in P. M. Harman, *The Natural Philosophy of James Clerk Maxwell* (Cambridge: Cambridge University Press, 1998), p. 171.

23. Quoted in John Theodore Merz, *A History of European Scientific Thought in the Nineteenth Century*, 2 vols. (New York: Dover, 1965), vol. 2, p. 62 n. 1, from Maxwell, *Scientific Papers*, vol. 2, p. 662.

24. Faraday quoted in Bruce J. Hunt, "Michael Faraday, Cable Telegraphy, and the Rise of Field Theory," *History of Technology* 13 (1991), pp. 1–19, on p. 4.

25. Quoted in ibid., pp. 14–15; also in Hunt, *The Maxwellians* (Ithaca: Cornell University Press, 1991), pp. 64–65.

26. Quoted in Smith and Wise, *Energy and Empire*, p. 386 (emphasis in original).

27. James Clerk Maxwell, "Atom," *Encyclopaedia Britannica*, 9th ed., vol. 3, pp. 36–49, on p. 49.

Chapter Six

1. Niels Bohr, *Collected Works*, ed. Léon Rosenfeld et al., 10 vols. (Amsterdam: North-Holland Publishing, 1972–99), vol. 2, p. 167.

2. A discussion from 1911 is Joseph Larmor, "Radiation," *Encyclopaedia Britannica*, 11th ed., 29 vols. (New York: Cambridge University Press, 1910–11), vol. 22, p. 792 col. 2.

3. Rutherford to Bohr, 20 March 1913, in Bohr, *Collected Works*, vol. 2, p. 583.

4. Quoted in Bohr, *Works*, vol. 2, p. 124.

5. Werner Heisenberg, "The Physical Content of Quantum Kinematics and Mechanics," in *Quantum Theory and Measurement*, ed. John Archibald Wheeler and Wojciech Hubert Zurek (Princeton: Princeton University Press, 1983), pp. 62–84, quote on p. 62 (trans. Wheeler and Zurek).

6. Ibid., p. 63; note that the translators generally render *anschaulich* as "physical."

7. Ibid., p. 83.

8. Niels Bohr, "The Quantum Postulate and the Recent Development of Atomic Theory," in Wheeler and Zurek, *Quantum Theory*, pp. 87–126, quote on p. 89.

9. Albert Einstein, "Physics and Reality" (first published 1936), in Einstein, *Ideas and Opinions* (New York: Crown Publishers, 1954), pp. 290–323, quote on p. 318.

10. Bohr, "The Quantum Postulate," p. 126.

11. James T. Cushing, *Quantum Mechanics: Historical Contingency and the Copenhagen Hegemony* (Chicago: University of Chicago Press, 1994), p. 25.

12. Mara Beller, *Quantum Dialogue: The Making of a Revolution* (Chicago: University of Chicago Press, 1999), p. 153.

13. Niels Bohr, "Can Quantum-Mechanical Description of Physical Reality Be Considered Complete?" in Wheeler and Zurek, *Quantum Theory*, pp. 145–51, quote on p. 150.

14. Ibid., p. 151.

15. Erwin Schrödinger, "The Present Situation in Quantum Mechanics," trans. John D. Trimmer, in Wheeler and Zurek, *Quantum Theory*, pp. 152–67, quote on p. 152.

16. Ibid., p. 156.

17. Ibid.

18. Ibid., p. 157.

19. Ibid.

20. Ibid.

21. Ibid., p. 161.

22. Ibid., p. 166.

23. Ibid.

24. Ibid.

25. David Bohm, *Causality and Chance in Modern Physics* (New York: Harper Torchbooks, 1961 [1957]), p. 96.

26. Ibid., p. 103.

Conclusion

1. Quoted in James W. McAllister, *Beauty and Revolution in Science* (Ithaca: Cornell University Press, 1996), p. 15.

2. Fred L. Bookstein, "Biotech and the Watchdog Role of Universities," *Washington Post*, 30 July 2001, A15. My thanks to Adrian Johns for having drawn my attention to this article.

3. Quoted in Stephen Clucas, "Thomas Harriot and the Field of Knowledge in the English Renaissance," in Robert Fox, ed., *Thomas Harriot: An Eliza-*

bethan Man of Science (Aldershot, UK: Ashgate, 2000), pp. 93–136, on p. 108 (spelling here modernized and Clucas's transcription emendations silently incorporated).

4. John D. Barrow and Frank J. Tipler, *The Anthropic Cosmological Principle* (Oxford: Clarendon Press, 1986), p. 2.

5. James Hutton, "Theory of the Earth," *Transactions of the Royal Society of Edinburgh* 1 (1788), pt. 2, pp. 209–304, on p. 304.

6. James Hutton, *Theory of the Earth with Proofs and Illustrations,* 2 vols. (Edinburgh, 1795), vol. 1, p. 3.

7. Ibid., p. 4.

8. Euclid, *The Thirteen Books of Euclid's Elements,* trans. Thomas L. Heath, 3 vols. (New York: Dover, 1956), vol. 1, p. 251.

9. Quoted in Simon Singh, *Fermat's Last Theorem* (London: Fourth Estate, 1997), p. 279.

10. Ian Stewart, quoted in Donald A. MacKenzie, *Mechanizing Proof: Computing, Risk, and Trust* (Cambridge, MA: MIT Press, 2001), p. 138.

11. Paul Teller, quoted in ibid., p. 143.

12. Steven G. Krantz, quoted in ibid., pp. 145, 149.

13. Interview material in ibid., p. 138.

Bibliographical Essay

This essay details scholarly literature most relevant to the various chapters of this book, as well as material that readers may find of interest in pursuing some of the issues further.

Introduction

The special and privileged place of the scientific and technical community in the administrative and legal operations of modern political states is an issue discussed by many scholars; among the books of particular interest are Sheila Jasanoff, *The Fifth Branch: Science Advisers as Policymakers* (Cambridge, MA: Harvard University Press, 1990); Jasanoff, *Science at the Bar: Law, Science, and Technology in America* (Cambridge, MA: Harvard University Press, 1995); Stephen Hilgartner, *Science on Stage: Expert Advice as Public Drama* (Stanford, CA: Stanford University Press, 2000); Dorothy Nelkin, ed., *The Politics of Technical Decisions,* 2nd ed. (Beverly Hills: Sage, 1984).

The belief that technical, operational effectiveness is the direct consequence of the truth of "pure" scientific theories is investigated and contested by Michael J. Mulkay, "Knowledge and Utility: Implications for the Sociology of Knowledge," *Social Studies of Science* 9 (1979), pp. 63–80, while discussions of similar issues that describe the two relevant dimensions in the terms of "science" and "technology" are Edwin T. Layton, Jr., "Conditions of Technological Development," in *Science, Technology and Society: A Cross-Disciplinary Perspective*, ed. Ina Spiegel-Rösing and Derek de Solla Price (London and Beverly Hills: Sage, 1977), pp. 197–222, Layton, "Through the Looking Glass, or News from Lake Mirror Image," *Technology and Culture* 28 (1987), pp. 594–607, among other writings by Layton; also Ronald L. Kline, "Construing 'Technology' as 'Applied Science': Public Rhetoric of Scientists and Engineers in the United States, 1880–1945," *Isis* 89 (1995), pp. 194–221; David Edgerton, "'The Linear Model' Did Not Exist: Reflections on the History and Historiography of Science and Research in Industry in the Twentieth Century," in *The Science-Industry Nexus: History, Policy, Implications*, ed. Karl Grandin, Nina Wormbs, and Sven Widmalm (Canton, MA: Science History Publications/USA, 2004), pp. 31–57. Even more counterintuitively, a landmark study in the 1960s by the U.S. government examined the extent to which practical achievements in defense-related research stemmed from "basic" research in the sciences, concluding that investment in basic science could not be said to have paid off in any evident way: Office of the Director of Defense Research and Engineering, *Project Hindsight—Final Report* (Washington: National Technical Information Service, 1967).

The observation that kinds of entities and characterizations of the contents of nature (atoms, light particles, the aether, etc.) have gone in and out of favor during the history of science was made most famously in Thomas S. Kuhn, *The Structure of Scientific Revolutions*, 2nd ed. (Chicago: University of Chicago Press, 1970), e.g., pp. 11–12, 111–17; also the interesting discussion of Kuhn's use of history in Isabelle Stengers, *The Invention of Modern Science*, trans. Daniel W. Smith (Minneapolis: University of Minnesota Press, 2000), chap. 3. The concept of "ideal types," applicable to all kinds of social institutions, is due to the early-twentieth-century German sociologist Max Weber: see *Max Weber on the Methodology of the Social Sciences*, ed. and trans. Edward A. Shils and Henry A. Finch (Glencoe, IL: Free Press, 1949). Gestalt psychology, represented by images that can be seen in several mutually exclusive ways, is classically employed in relation to science by Norwood Russell Hanson, *Patterns of Discovery: An Inquiry into the Conceptual Foundations of Science* (Cambridge: Cambridge University Press, 1958), pp. 11–15, but to make a point rather different from the one involved here.

Arguments by philosophers of science that hold the practical successes of science as evidence for the (approximate) truth of scientific theories most often come from those holding the position known as "reference realism." A classic statement of that position is Richard N. Boyd, "The Current Status of Scientific Realism," in *Scientific Realism*, ed. Jarrett Leplin (Berkeley: University of California Press, 1984), pp. 41–82; a recent on-line discussion is Richard Boyd, "Scientific Realism," in *The Stanford Encyclopedia of Philosophy* (summer 2002 ed.), ed. Edward N. Zalta, http://plato.stanford.edu/archives/sum2002/entries/scientific-realism/.

In addition to the references by Mulkay and others cited above, further arguments in support of the view that the practical "application" of scientific theory involves a great deal of additional practical and intellectual work may be found in Andrew Pickering, *The Mangle of Practice: Time, Agency, and Science* (Chicago: University of Chicago Press, 1995).

On the classical Greek concepts of *epistēmē* and *technē*, see the accessible account in Pamela O. Long, *Openness, Secrecy, Authorship: Technical Arts and the Culture of Knowledge from Antiquity to the Renaissance* (Baltimore: Johns Hopkins University Press, 2001), esp. p. 2 (for definitions), chap. 1, and epilogue for a summary of the book's more extended argument; see also Peter Dear, "The Ideology of Modern Science" (essay-review of Long, *Openness, Secrecy, Authorship*), *Studies in History and Philosophy of Science* 34A (2003), pp. 821–28, for a somewhat more detailed version of my argument concerning natural philosophy and instrumentality, and, at greater length, Dear, "What Is the History of Science the History *Of*? Early Modern Roots of the Ideology of Modern Science," *Isis* 96 (2005), pp. 390–406.

G. E. R. Lloyd is arguably the best contemporary scholar of ancient Greek science; of his many books, see esp. G. E. R. Lloyd, *Magic, Reason, and Experience: Studies in the Origin and Development of Greek*

Science (Cambridge: Cambridge University Press, 1979). A brief account of the situation during the European Middle Ages and early-modern period may be found in Peter Dear, *Revolutionizing the Sciences: European Knowledge and Its Ambitions, 1500–1700* (Princeton: Princeton University Press, 2001), introduction and chap. 1, while Pamela H. Smith, *The Body of the Artisan: Art and Experience in the Scientific Revolution* (Chicago: University of Chicago Press, 2004), presents a detailed argument for the importance of practical artisanal engagement with nature. The development of the Newtonian ideology in early eighteenth-century Britain is detailed in Larry Stewart, *The Rise of Public Science: Rhetoric, Technology, and Natural Philosophy in Newtonian Britain, 1660–1750* (Cambridge: Cambridge University Press, 1992); some opposition to Newton's innovations is mentioned in Peter Dear, *Discipline and Experience: The Mathematical Way in the Scientific Revolution* (Chicago: University of Chicago Press, 1995), conclusion, as well as in Dear, *Revolutionizing the Sciences*, pp. 164–67.

The label "technoscience," signifying a lack of real distinction between "science" and "technology," has been made popular by Bruno Latour: Bruno Latour, *Science in Action: How to Follow Scientists and Engineers through Society* (Cambridge, MA: Harvard University Press, 1987), esp. pp. 174–75. It has been taken up, for example, in the domain of cultural studies, e.g., George Marcus, ed., *Technoscientific Imaginaries: Conversations, Profiles, and Memoirs* (Chicago: University of Chicago Press, 1995). The view of science carried in the word "technoscience," however, is one directly opposed to the perspective suggested here, insofar as the term seems to deny any fundamental distinction between what I call natural philosophy and instrumentality. "Technoscience" implies that science (in my terms, a hybrid) and technology (concerned above all with instrumentality) are essentially the same thing. For a fascinating case study, see Adelheid Voskuhl, "Humans, Machines, and Conversations: An Ethnographic Study of the Making of Automatic Speech Recognition Technologies," *Social Studies of Science* 34 (2004), pp. 393–421.

The key notion of "intelligibility" is discussed at greater length in Peter Dear, "Intelligibility in Science," *Configurations* 11 (2003), pp. 145–

61. The idea relies on a clear distinction between "explanation" (a formal process, often deductive), and "understanding," a fundamentally nonformalized relationship between a knower and an explanatory scheme: see for explication James T. Cushing, *Quantum Mechanics: Historical Contingency and the Copenhagen Hegemony* (Chicago: University of Chicago Press, 1994), pp. 10–12. Howard E. Gruber, "On the Relation between 'Aha Experiences' and the Construction of Ideas," *History of Science* 19 (1981), pp. 41–59, is an interesting study of the "flash of insight," a psychologically related aspect of intelligibility. Leonard K. Nash, *The Nature of the Natural Sciences* (Boston: Little, Brown, 1963), used what he called the "principle of intelligibility" as one of the essential hallmarks of science, but regarded it as essentially unproblematic and timeless; see esp. pp. 65–66, 172–76.

Chapter One

For a general introduction to the issues of this chapter, see Dear, *Revolutionizing the Sciences;* John Henry, *The Scientific Revolution and the Origins of Modern Science,* 2nd ed. (Basingstoke: Palgrave, 2002). Marie Boas, "The Establishment of the Mechanical Philosophy," *Osiris* 10 (1952), pp. 412–541, set the form of a great deal of subsequent literature on the "mechanical philosophy" down to the present day. On Descartes, two excellent treatments are Stephen Gaukroger, *Descartes: An Intellectual Biography* (Oxford: Clarendon Press, 1995); William R. Shea, *The Magic of Numbers and Motion: The Scientific Career of René Descartes* (New York: Science History Publications, 1991), the latter focusing especially on Descartes's technical scientific work. For a philosophical discussion of Descartes's mechanism, see Daniel Garber, *Descartes' Metaphysical Physics* (Chicago: University of Chicago Press, 1992).

Seventeenth-century criticisms of scholastic Aristotelian explanations are discussed and debated by Keith Hutchison, "Dormitive Virtues, Scholastic Qualities, and the New Philosophies," *History of Science* 29 (1991), pp. 245–78, and Desmond Clarke, "Dormitive Powers and Scholastic Qualities: A Reply to Hutchison," *History of Science* 31 (1993), pp. 317–24; see also Hutchison, "What Happened to Occult Qualities in the Scientific Revolution?," *Isis* 73 (1982), pp. 233–53. For more comprehensive analysis and review of Aristotelian arguments and their critics, see Dennis Des Chene, *Physiologia: Natural Philosophy in Late Aristotelian and Cartesian Thought* (Ithaca: Cornell University Press, 1996), while the same author's *Spirits and Clocks: Machine and Organism in Descartes* (Ithaca: Cornell University Press, 2001) looks closely at Descartes's own use of the machine image in understanding living things. A study of this latter theme consistent with the arguments of the present book is Peter Dear, "A Mechanical Microcosm: Bodily Passions, Good Manners, and Cartesian Mechanism," in *Science Incarnate: Historical Embodiments of Natural Knowledge*, ed. Christopher Lawrence and Steven Shapin (Chicago: University of Chicago Press, 1998), pp. 51–82.

Robert Kargon, *Atomism in England from Hariot to Newton* (Oxford: Clarendon Press, 1966), is a classic account of corpuscular mechanism (Boyle's "mechanical philosophy") in England in this period; however, an important article that holds that much seventeenth-century English "mechanical philosophy" deviated a great deal from the ideal of inert matter in motion promoted by Descartes is John Henry, "Occult Qualities and the Experimental Philosophy: Active Principles in Pre-Newtonian Matter Theory," *History of Science* 24 (1986), pp. 335–81; on Newton's own ideas about "active principles" that would supplement the inert properties of mechanically understood matter, see the classic accounts by J. E. McGuire, "Force, Active Principles, and Newton's Invisible Realm," *Ambix* 15 (1968), pp. 154–208, and P. M. Heimann, "'Nature is a Perpetual Worker': Newton's Aether and Eighteenth-Century Natural Philosophy," *Ambix* 20 (1973), pp. 1–25, and, more recently, Alan Gabbey, "Newton, Active Powers, and the Mechanical Philosophy," in *The Cambridge Companion to Newton*, ed. I. Bernard Cohen and George E. Smith (Cambridge: Cambridge University Press, 2002), pp. 329–57.

For more on Boyle's mechanism, see Peter Alexander, *Ideas, Qualities and Corpuscles: Locke and Boyle on the External World* (Cambridge:

Cambridge University Press, 1985), and, on Boyle's natural philosophy more generally, Rose-Mary Sargent, *The Diffident Naturalist: Robert Boyle and the Philosophy of Experiment* (Chicago: University of Chicago Press, 1995), and Peter R. Anstey, *The Philosophy of Robert Boyle* (London: Routledge, 2000). A contrast between the approaches to natural philosophy of Boyle and of Hobbes (focusing, however, on the issue of experiment rather than on their approaches to the philosophy of mechanism) is the subject of Steven Shapin and Simon Schaffer, *Leviathan and the Air-Pump: Hobbes, Boyle, and the Experimental Life* (Princeton: Princeton University Press, 1985), which illuminatingly discusses, inter alia, Boyle's idea of the "spring of the air." Hobbes's approach to natural philosophy, which he based on the deductive reasoning of geometry, is usefully discussed by Douglas M. Jesseph, *Squaring the Circle: The War between Hobbes and Wallis* (Chicago: University of Chicago Press, 1999); also Jesseph, "Hobbes and the Method of Natural Science," in *The Cambridge Companion to Hobbes,* ed. Tom Sorrell (Cambridge: Cambridge University Press, 1996), pp. 86–107.

On Huygens and hypotheses, good places to start are Aant Elzinga, *On a Research Program in Early Modern Physics* (Göteborg: Akademiförlaget, 1972); Robert S. Westman, "Huygens and the Problem of Cartesianism," in *Studies on Christiaan Huygens: Invited Papers from the Symposium on the Life and Work of Christiaan Huygens, Amsterdam, 22–25 August 1979,* ed. H. J. M. Bos, et al. (Lisse: Swets & Zeitlinger, 1980), pp. 83–103; Roberto De A. Martins, "Huygens's Reaction to Newton's Gravitational Theory," in *Renaissance and Revolution: Humanists, Scholars, Craftsmen and Natural Philosophers in Early Modern Europe,* ed. J. V. Field and Frank A. J. L. James (Cambridge: Cambridge University Press, 1993), pp. 203–13. The standard biography of Newton is Richard S. Westfall, *Never at Rest: A Biography of Isaac Newton* (Cambridge: Cambridge University Press, 1980); on Newton's methodological and metaphysical views and on those of some of his critics, see the classic account in Alexandre Koyré, *From the Closed World to the Infinite Universe* (Baltimore: Johns Hopkins University Press, 1957), chap. 7. See more generally Gerd Buchdahl, "Gravity and In-

telligibility: Newton to Kant," in *The Methodological Heritage of Newton,*
ed. Robert E. Butts and John W. Davis (Toronto: University of
Toronto Press, 1968), pp. 74–102; for a historical survey see Peter
Dear, "Method and the Study of Nature," in *The Cambridge History of
Seventeenth-Century Philosophy,* ed. Daniel Garber and Michael Ayers
(Cambridge: Cambridge University Press, 1998), pp. 147–77, esp.
pp. 161–70. The reception of Newton's optical arguments is discussed
in Simon Schaffer, "Glass Works: Newton's Prisms and the Uses of
Experiment," in *The Uses of Experiment: Studies in the Natural Sciences,*
ed. Trevor Pinch, Simon Schaffer and David Gooding (Cambridge:
Cambridge University Press, 1989), pp. 67–104. On the role of God
in early-modern natural philosophy, see Andrew Cunningham,
"How the *Principia* Got Its Name; or, Taking Natural Philosophy Seriously," *History of Science* 29 (1991), pp. 377–92.

   Much the most important study for the topic of this chapter is
Alan Gabbey, "The Mechanical Philosophy and Its Problems: Me-
chanical Explanations, Impenetrability, and Perpetual Motion," in
Change and Progress in Modern Science, ed. Joseph C. Pitt (Dordrecht: D.
Reidel, 1985), pp. 9–84, which shows in detail the difficulties recog-
nized by contemporaries themselves with the foundational concepts
of mechanical explanation. See also Gabbey, "Henry More and the
Limits of Mechanism," in *Henry More (1614–1687): Tercentenary Studies,*
ed. Sarah Hutton (Dordrecht: Kluwer, 1990), pp. 19–35; Gabbey,
"Newton's *Mathematical Principles of Natural Philosophy:* A Treatise on
'Mechanics'?" in *The Investigation of Difficult Things: Essays on Newton
and the History of the Exact Sciences,* ed. Peter Harman and Alan Shapiro
(Cambridge: Cambridge University Press, 1992), pp. 305–22.

Chapter Two

Opponents of the mechanistic idea that living beings were just elaborate machines (Descartes's view) were quite plentiful in the eighteenth century. Among them were the Newtonian natural philosopher and mathematician Maupertuis and the great naturalist Buffon:

see Mary Terrall, *The Man Who Flattened the Earth: Maupertuis and the Sciences in the Enlightenment* (Chicago: University of Chicago Press, 2002), chap. 7; Jacques Roger, *Buffon: A Life in Natural History*, trans. Sarah Lucille Bonnefoi (Ithaca: Cornell University Press, 1997), esp. chaps. 9, 18.

On tables in eighteenth-century chemistry, Mi Gyung Kim, *Affinity, That Elusive Dream: A Genealogy of the Chemical Revolution* (Cambridge, MA: MIT Press, 2003); Alistair Duncan, *Laws and Order in Eighteenth-Century Chemistry* (Oxford: Clarendon Press, 1996); Arnold Thackray, "Quantified Chemistry: The Newtonian Dream," in *John Dalton and the Progress of Science*, ed. D. S. L. Cardwell (Manchester: Manchester University Press, 1968), pp. 92–108; Lissa Roberts, "Setting the Table: The Disciplinary Development of Eighteenth-Century Chemistry as Read through the Changing Structure of Its Tables," in *The Literary Structure of Scientific Argument: Historical Studies*, ed. Peter Dear (Philadelphia: University of Pennsylvania Press, 1991), pp. 99–132. Geoffroy's chemical career is surveyed in W. A. Smeaton, "Geoffroy, Étienne François," in *Dictionary of Scientific Biography*, ed. Charles Coulston Gillispie (New York: Scribner's, 1970–80), vol. 5, pp. 352–54; see also Frederic L. Holmes, *Eighteenth-Century Chemistry as an Investigative Enterprise* (Berkeley: Office for History of Science and Technology, University of California at Berkeley, 1989), chap. 2; Holmes, "The Communal Context for Etienne-François Geoffroy's 'Table des rapports,'" *Science in Context* 9 (1996), pp. 289–311 (followed by an English translation of Geoffroy's paper on pp. 313–20). Further contextual discussion appears in Mi Gyung Kim, "Chemical Analysis and the Domains of Reality: Wilhelm Homberg's *Essais de chimie*, 1702–1709," *Studies in History and Philosophy of Science* 31 (2000), pp. 37–69. On crystalline structure as a criterion in the classification of salts, see Holmes, *Eighteenth-Century Chemistry*, p. 52.

John E. Lesch, "Systematics and the Geometrical Spirit," in *The Quantifying Spirit in the Eighteenth Century*, ed. Tore Frängsmyr, J. L. Heilbron, and Robin E. Rider (Berkeley: University of California Press, 1990), pp. 73–111, is a valuable overview of classificatory endeavors in this period in a wide variety of fields. Lesch suggests a general similarity between such systematics and the equally wide-spread Newtonian mathematizing ambition, which also sought codifiable order in the natural world.

For an introduction to classification in eighteenth-century botany and zoology, see Paul Lawrence Farber, *Finding Order in Nature: The Naturalist Tradition from Linnaeus to E. O. Wilson* (Baltimore: Johns Hopkins University Press, 2000), chaps. 1–3. On some of the crucial epistemological controversies among systematists of the period, see esp. Phillip R. Sloan, "John Locke, John Ray, and the Problem of the Natural System," *Journal of the History of Biology* 5 (1972), pp. 1–53; on Buffon, in addition to Roger, *Buffon*, see Sloan, "The Buffon-Linnaeus Controversy," *Isis* 67 (1976), pp. 356–75; Sloan, "Buffon, German Biology, and the Historical Interpretation of Biological Species," *British Journal for the History of Science* 12 (1979), pp. 109–53, as well as the persuasive revisionist account by John H. Eddy, Jr., "Buffon's *Histoire naturelle:* History? A Critique of Recent Interpretations," *Isis* 85 (1994), pp. 644–61. A good source of primary texts in translation is John Lyon and Phillip R. Sloan, eds., *From Natural History to the History of Nature: Readings from Buffon and His Critics* (Notre Dame: University of Notre Dame Press, 1981). For the institutional structure of French natural history in the period, see E. C. Spary, *Utopia's Garden: French Natural History from Old Regime to Revolution* (Chicago: University of Chicago Press, 2000). An interesting discussion of a much later dispute concerning biological classification is John Dean, "Controversy over Classification: A Case Study from the History of Botany," in *Natural Order: Historical Studies of Scientific Culture*, ed. Barry Barnes and Steven Shapin (Beverly Hills: Sage, 1979), pp. 211–30.

An increasingly recognized, and important, context for natural history in the eighteenth century is that of European imperialism and colonialism, a context that becomes increasingly evident in the nineteenth century. Especially important is Lisbet Koerner, *Linnaeus: Nature and Nation* (Cambridge, MA: Harvard University Press, 1999); a useful overview of similar issues is Lisbet Koerner, "Carl Linnaeus in His Time and Place," in *Cultures of Natural History*, ed. Nicholas Jardine, James A. Secord, and Emma C. Spary (Cambridge: Cambridge University Press, 1996), pp. 145–62. Natural history as an integral part of the European voyages of discovery is discussed in John Gascoigne, *Science in the Service of Empire: Joseph Banks, the British State and the Uses of Science in the Age of Revolution* (Cambridge: Cambridge University Press, 1998), as well as essays in David Philip Miller and Peter Hanns Reill, eds., *Visions of Empire: Voyages, Botany, and Representations of Nature* (Cambridge: Cambridge University Press, 1996); a short popular treatment of these issues is Patricia Fara, *Sex, Botany, and Empire: The Story of Carl Linnaeus and Joseph Banks* (New York: Columbia University Press, 2004), while a particularly fascinating study is Kapil Raj, "Surgeons, Fakirs, Merchants, and Craftspeople: Making L'Empereur's *Jardin* in Early Modern South Asia," in *Colonial Botany: Science, Commerce, and Politics in the Early Modern World*, ed. Claudia Swan and Londa Schiebinger (Philadelphia: University of Pennsylvania Press, 2005), pp. 252–69. Another interesting aspect of Linnaeus's system is discussed in Londa Schiebinger, "Why Mammals Are Called Mammals," in Schiebinger, *Nature's Body: Gender in the Making of Modern Science* (Boston: Beacon Press, 1993), chap. 2.

On Ramus and the role of his pedagogical theories in textbooks, an excellent treatment relating to seventeenth-century chemistry is Owen Hannaway, *The Chemists and the Word: The Didactic Origins of Chemistry* (Baltimore: Johns Hopkins University Press, 1975); Hannaway's lead is followed into the eighteenth century in John R. R. Christie and J. V. Golinski, "The Spreading of the Word: New Directions in the Historiography of Chemistry, 1600–1800," *History of Science* 20 (1982), pp. 235–66. For Haüy and mineralogy, see Lesch, "Systematics," pp. 88–92; also Peter F. Stevens, "Haüy and A.-P. Candolle: Crystallography, Botanical Systematics, and Comparative Morphology," *Journal of the History of Biology* 17 (1984), pp. 49–82.

A classic account of eighteenth-century ideas about the hierarchical relatedness of organic beings is Arthur O. Lovejoy, *The Great Chain of Being: A Study of the History of an Idea* (Cambridge, MA: Harvard University Press, 1936), while Michel Foucault, *The Order of Things: An Archaeology of the Human Sciences* (New York: Vintage Books, 1973), chap. 5, is a seminal discussion of the meaning of ordering the world at this time. A valuable article strongly influenced by Foucault's conceptions is Simon Schaffer, "Herschel in Bedlam: Natural History and Stellar Astronomy," *British Journal for the History of Science* 13 (1981), pp. 211–39. For more material on Herschel, see also M. A. Hoskin, *William Herschel and the Construction of the Heavens* (London: Oldbourne, 1963).

In later eighteenth-century Prussia, the philosophical proposals of Immanuel Kant also began to reconceive the category of "natural history," but the most important practical taxonomic departures in the period were still French; for a broad look at these issues, see Nicholas Jardine, "Inner History; or, How to End Enlightenment," in *The Sciences in Enlightened Europe*, ed. William Clark, Jan Golinski, and Simon Schaffer (Chicago: University of Chicago Press, 1999), pp. 477–94. On Cuvier's ideas, see Martin J. S. Rudwick, *Georges Cuvier, Fossil Bones, and Geological Catastrophes: New Translations and Interpretations of the Primary Texts* (Chicago: University of Chicago Press, 1997); also Martin J. S. Rudwick, *The Meaning of Fossils: Episodes in the History of Palaeontology*, 2nd ed. (Chicago: University of Chicago Press, 1985), chap. 3; William Coleman, *Georges Cuvier, Zoologist: A Study in the History of Evolution Theory* (Cambridge, MA: Harvard University Press, 1964); Dorinda Outram, "Uncertain Legislator: Georges Cuvier's Laws of Nature in Their Intellectual Context," *Journal of the History of Biology* 19 (1986), pp. 323–68.

Finally, an important consideration of classification as a form of knowledge making, and the complexity of the ways in which it is achieved, is Barry Barnes, "On the Conventional Character of Knowledge and Cognition," in *Science Observed: Perspectives on the Social Study of Science*, ed. Karin D. Knorr-Cetina and Michael J. Mulkay (London: Sage, 1983), pp. 19–51. See also, for a discussion of the implications of classificatory systems, Geoffrey C. Bowker and Susan Leigh Star, *Sorting Things Out: Classification and Its Consequences* (Cambridge, MA: MIT Press, 1999).

Chapter Three

On early-modern chemical pedagogy, the classic study is Hannaway, *The Chemists and the Word*. Hannaway's themes are developed and expanded by Christie and Golinski, "The Spreading of the Word." On the alchemical perspective, see the recent study by William R. Newman and Lawrence M. Principe, *Alchemy Tried in the Fire: Starkey, Boyle, and the Fate of Helmontian Chymistry* (Chicago: University of Chicago Press, 2002).

J. B. Gough, "Lavoisier and the Fulfillment of the Stahlian Revolution," *Osiris* n.s. 4 (1988), pp. 15–33, surveys some aspects of the relationship between Stahl's and Lavoisier's ideas. For discussion of Stahl's chemical views, see Arnold Thackray, *Atoms and Powers: An Essay on Newtonian Matter-Theory and the Development of Chemistry* (Cambridge, MA: Harvard University Press, 1970), pp. 171–76; the book also considers many other issues of chemistry in relation to matter theory in the eighteenth century. See also Duncan, *Laws and Order*, pp. 52–56, and Kim, *Affinity*, pp. 168–74. Stahl's English translator was Peter Shaw, on whom see J. V. Golinski, "Peter Shaw: Chemistry and Communication in Augustan England," *Ambix* 30 (1983), pp. 19–29. The foundational twentieth-century study behind all subsequent historical literature on this topic is Hélène Metzger, *Newton, Stahl, Boerhaave et la doctrine chimique* (Paris: Alcan, 1930).

On the use of the senses by chemists in this period, see Lissa Roberts, "The Death of the Sensuous Chemist: The 'New' Chemistry and the Transformation of Sensuous Technology," *Studies in History and Philosophy of Science* 26 (1995), pp. 503–29. Venel is an important figure in Charles Coulston Gillispie, "The *Encyclopédie* and the Jacobin Philosophy of Science: A Study in Ideas and Consequences," in *Critical Problems in the History of Science*, ed. Marshall Clagett (Madison: University of Wisconsin Press, 1959), pp. 255–89.

Berthollet's attempts at the beginning of the nineteenth century to understand affinity tables in terms of the forces acting between the particles of matter are discussed in Kim, *Affinity*, pp. 417–38; see also Maurice P. Crosland, *The Society of Arcueil: A View of French Science at the Time of Napoleon* (Cambridge, MA: Harvard University Press, 1967), and Robert Fox, "The Rise and Fall of Laplacian Physics," *Historical Studies in the Physical Sciences* 4 (1974), pp. 89–136, a shorter version of which appears as Fox, "Laplacian Physics," in *Companion to the History of Modern Science,* ed. R. C. Olby et al. (Boston: Routledge, 1990), pp. 278–394.

A good general introduction to Lavoisier and his work is William H. Brock, *The Norton History of Chemistry* (New York: Norton, 1993), chap. 3; see also Kim, *Affinity,* chap. 6. An excellent biographical treatment is Arthur Donovan, *Antoine Lavoisier: Science, Administration, and Revolution* (Cambridge: Cambridge University Press, 1996); see esp. chap. 7 on the new chemical nomenclature.

On Lavoisier's claims to experimental precision and accuracy in his chemical work, as well as British reactions to him, see Jan Golinski, "'The Nicety of Experiment': Precision of Measurement and Precision of Reasoning in Late Eighteenth-Century Chemistry," in *The Values of Precision,* ed. M. Norton Wise (Princeton: Princeton University Press, 1995), pp. 72–91, as well as Jan Golinski, *Science as Public Culture: Chemistry and Enlightenment in Britain, 1760–1820* (Cambridge: Cambridge University Press, 1992), chap. 5. The general issue of claims to mathematical precision by French experimentalists in the late eighteenth century is considered in Christian Licoppe, *La formation de la pratique scientifique: Le discours de l'expérience en France et en Angleterre, 1630–1820* (Paris: Éditions La Découverte, 1996), chap. 7.

The standard biography of John Dalton is Frank Greenaway, *John Dalton and the Atom* (Ithaca: Cornell University Press, 1966). Considerations of Dalton's chemical ideas, together with primary documents, appear in Arnold Thackray, *John Dalton: Critical Assessments of His Life and Science* (Cambridge, MA: Harvard University Press, 1972), while D. S. L. Cardwell, ed., *John Dalton and the Progress of Science* (Manchester: Manchester University Press, 1968), is a valuable collection of essays on Dalton and related aspects of eighteenth-century chemistry; see also Thackray, *Atoms and Powers,* pp. 252–78. For Dalton's reception, see L. A. Whitt, "Atoms or Affinities? The Ambivalent Reception of Daltonian Theory," *Studies in History and*

Philosophy of Science 21 (1990), pp. 57–89. Newton's discussions of the nature of gases may be found in Isaac Newton, *The Principia—Mathematical Principles of Natural Philosophy: A New Translation and Guide*, trans. I. Bernard Cohen and Anne Whitman (Berkeley: University of California Press, 1999), bk. 2, prop. 23; Newton, *Opticks, or A Treatise of the Reflections, Refractions, Inflections & Colours of Light* (New York: Dover, 1952), qu. 31 (esp. pp. 395–96). On the subsequent career of chemical atomism in the nineteenth century, see Alan J. Rocke, *Chemical Atomism in the Nineteenth Century: From Dalton to Cannizzaro* (Columbus: Ohio State University Press, 1984); Mary Jo Nye, *Before Big Science: The Pursuit of Modern Chemistry and Physics 1800–1940* (Cambridge, MA: Harvard University Press, 1999), chap. 2.

Chapter Four

The most pleasing edition still in print of this chapter's fundamental, and endlessly rewarding, book is Charles Darwin, *On the Origin of Species*, facsimile of the 1st ed., ed. Ernst Mayr (Cambridge, MA: Harvard University Press, 1964); a useful recent edition with supplementary materials is Charles Darwin, *On the Origin of Species by Means of Natural Selection*, ed. Joseph Carroll (Peterborough, Ontario: Broadview, 2003).

There are many biographies of Darwin, several of them outstanding in quality. The most recent, and arguably the best, is the two-volume treatment by Janet Browne: vol. 1, *Charles Darwin: Voyaging* (Princeton: Princeton University Press, 1995), vol. 2, *Charles Darwin: The Power of Place* (Princeton: Princeton University Press, 2002).

On William Paley, natural theology, and their role in Darwin's thought, see the classic studies by Dov Ospovat, "Perfect Adaptation and Teleological Explanation: Approaches to the Problem of the History of Life in the Mid-nineteenth Century," *Studies in the History of Biology* 2 (1978), pp. 33–56, and *The Development of Darwin's Theory: Natural History, Natural Theology, and Natural Selection, 1838–1859* (Cambridge: Cambridge University Press, 1981); Neal C. Gillespie, "Divine Design and the Industrial Revolution: William Paley's Abortive Reform of Natural Theology," *Isis* 81 (1990), pp. 214–29; on the career of Paley's book, see Aileen Fyfe, "Publishing and the Classics: Paley's *Natural Theology* and the Nineteenth-Century Scientific Canon," *Studies in History and Philosophy of Science* 33A (2002), pp. 729–51. See also the useful account in John Hedley Brooke, *Science and Religion: Some Historical Perspectives* (Cambridge: Cambridge University Press, 1991), which looks as well at parts of this British tradition prior to the nineteenth century.

On classification in the nineteenth century, see Harriet Ritvo, *The Platypus and the Mermaid, and Other Figments of the Classifying Imagination* (Cambridge, MA: Harvard University Press, 1997). Mary P. Winsor, *Starfish, Jellyfish, and the Order of Life: Issues in Nineteenth-Century Science* (New Haven: Yale University Press, 1976), chap. 8, discusses Darwin's conception of a changing, rather than static, taxonomical system resulting from transmutations over time that create new filiations or obliterate old ones. On Darwin and domestic breeding see James A. Secord, "Nature's Fancy: Charles Darwin and the Breeding of Pigeons," *Isis* 72 (1981), pp. 163–86, and Secord, "Darwin and the Breeders: A Social History," in *The Darwinian Heritage*, ed. David Kohn (Princeton: Princeton University Press, 1985), pp. 519–42.

The challenge of William Thomson, Lord Kelvin, to the idea of an immensely old earth is the subject of Joe D. Burchfield, *Lord Kelvin and the Age of the Earth* (New York: Science History Publications, 1975); also Crosbie Smith and M. Norton Wise, *Energy and Empire: A Biographical Study of Lord Kelvin* (Cambridge: Cambridge University Press, 1989), chaps. 16, 17.

The work of Gillian Beer is especially important for discussions of the literary dimensions of Darwin's *Origin:* Gillian Beer, *Darwin's Plots: Evolutionary Narratives in Darwin, George Eliot, and Nineteenth-Century Fiction* (London: Routledge & Kegan Paul, 1983); essays in section 1 of Beer, *Open Fields: Science in Cultural Encounter* (Oxford: Oxford University Press, 1996). See also the classic article by Robert M. Young, "Darwin's Metaphor: Does Nature Select?" in *Darwin's Metaphor: Nature's Place in Victorian Culture* (Cambridge: Cambridge University

Press, 1985), chap. 4, which examines the resonances of some of Darwin's arguments in Victorian Britain. The huge expanses of time to which Darwin appealed in his arguments derived in large part from the work of contemporary geologists, of whom Darwin was himself a notable example in his earlier career. On nineteenth-century representations of the prehuman past, see Martin J. S. Rudwick, *Scenes from Deep Time: Early Pictorial Representations of the Prehistoric World* (Chicago: University of Chicago Press, 1992).

J. Vernon Jensen, "Return to the Wilberforce-Huxley Debate," *British Journal for the History of Science* 21 (1988), pp. 161–79, usefully investigates the famous encounter between the two men and plausibly suggests that the later story was elaborated to inflate Huxley's image as the fearless champion of scientific truth. Huxley's scientific work is studied in Mario A. Di Gregorio, *T. H. Huxley's Place in Natural Science* (New Haven: Yale University Press, 1984), with Huxley's concept of saltations discussed on pp. 65–68; see also Adrian Desmond, *Archetypes and Ancestors: Palaeontology in Victorian London 1850–1875* (Chicago: University of Chicago Press, 1984), which focuses on the rivalry between Huxley and Richard Owen. For Huxley's criticisms of Cuvier's functional approach to comparative anatomy, see Sherrie L. Lyons, *Thomas Henry Huxley: The Evolution of a Scientist* (Amherst, NY: Prometheus Books, 1999), pp. 55–60; also Adrian Desmond, *Huxley: The Devil's Disciple* (London: Michael Joseph, 1994), p. 227.

A useful overview of evolution and the "evolutionary synthesis" in the later nineteenth and earlier twentieth centuries is William B. Provine's *The Origins of Theoretical Population Genetics*, 2nd ed. (Chicago: University of Chicago Press, 2001). A classic article expounding Stephen Jay Gould's dissatisfaction with pure adaptationist assumptions is S. J. Gould and R. C. Lewontin, "The Spandrels of San Marco and the Panglossian Paradigm: A Critique of the Adaptationist Programme," *Proceedings of the Royal Society of London*, series B, vol. 205, no. 1161 (1979), pp. 581–598. Michael J. Behe, *Darwin's Black Box: The Biochemical Challenge to Evolution* (New York: Touchstone, 1996), is a celebrated example of the recent "intelligent design" anti-natural-selection position; arguments for the implausibility of natural selection's sufficiency are central to this view, which is rejected by most biologists. A good recent account of evolution and its cultural meanings down to the present is Edward J. Larson, *Evolution: The Remarkable History of a Scientific Theory* (New York: Modern Library, 2004).

Chapter Five

The contrasting of different national styles in physics, with an emphasis on British "mechanicism," is a theme of Pierre Duhem, *The Aim and Structure of Physical Theory*, trans. Philip P. Wiener from French 2nd ed., 1914 (Princeton: Princeton University Press, 1954); a similar, but British, perspective from the same general period is John Theodore Merz, *A History of European Thought in the Nineteenth Century*, 4 vols. (London, 1904–12), vol. 2, chap. 6 (reprinted in Merz, *A History of European Scientific Thought in the Nineteenth Century*, 2 vols. [New York: Dover, 1965]).

On early nineteenth-century ideas of the luminiferous aether, see Jed Z. Buchwald, *The Rise of the Wave Theory of Light: Optical Theory and Experiment in the Early 19th Century* (Chicago: University of Chicago Press, 1989); E. T. Whittaker, *A History of the Theories of Aether and Electricity*, 2 vols. (revised ed., London: Nelson, 1951), vol. 1, esp. chap. 5 on early mathematical aethers after Fresnel. Thomson's vortex atoms are discussed in Helge Kragh, "The Vortex Atom: A Victorian Theory of Everything," *Centaurus* 44 (2002), pp. 32–114.

On William Thomson, Crosbie Smith and M. Norton Wise, *Energy and Empire: A Biographical Study of Lord Kelvin* (Cambridge: Cambridge University Press, 1989), is much the most important study, while Maxwell's natural philosophy is treated in P. M. Harman, *The Natural Philosophy of James Clerk Maxwell* (Cambridge: Cambridge University Press, 1998); John Hendry, *James Clerk Maxwell and the Theory of the Electromagnetic Field* (Boston: Hilger, 1986); Alan F. Chalmers, "The Heuristic Role of Maxwell's Mechanical Model of Electromagnetic Phenomena," *Studies in History and Philosophy of Science* 17 (1986), pp. 415–27; Daniel M. Siegel, *Innovation in Maxwell's Electromagnetic Theory: Molecular Vortices, Displacement Current, and Light* (Cambridge: Cambridge University Press, 1991); Bruce J. Hunt, *The Maxwellians* (Ithaca: Cornell University Press, 1991). See also, more generally, Enrico Bellone, *A World on Paper: Studies on the Second Scientific Revolution*, trans. Mirella and Ricardo Giacconi (Cambridge, MA: MIT Press, 1980), chap. 3, and Nye, *Before Big Science*, chap. 3. Jed Z. Buchwald, *From Maxwell to Microphysics: Aspects of Electromagnetic Theory in the Last Quarter of the Nineteenth Century* (Chicago: University of Chicago Press, 1985), pt. 1, chap. 3, esp. pp. 20–23, is particularly straightforward on the distinction between a specific mechanical model and a dynamical system for Maxwell; Buchwald also notes that the dynamical ideal was "a future hope of the era" rather than "a practical method of investigation" (p. 20); see, for another technical discussion along similar lines, Olivier Darrigol, *Electrodynamics from Ampère to Einstein* (Oxford: Oxford University Press, 2000), chap. 4.

For material on responses by German/Austrian physicists to the hypothetical mechanical foundationalism common in British physics (they tended to dislike it), see Christa Jungnickel and Russell McCormmach, *Intellectual Mastery of Nature: Theoretical Physics from Ohm to Einstein*, 2 vols. (Chicago: University of Chicago Press, 1986), vol. 2. Ludwig Boltzmann's adherence to mechanics as foundational in regard to thermodynamics (ironically, in this case contrary to Maxwell) is discussed in ibid., pp. 154–57, and in Martin J. Klein, "Mechanical Explanation at the End of the Nineteenth Century," *Centaurus* 17 (1972), pp. 58–82. Also, for British physics c. 1900, see Andrew Warwick, *Masters of Theory: Cambridge and the Rise of Mathematical Physics* (Chicago: University of Chicago Press, 2003), esp. chaps. 6, 7.

The association between nineteenth-century industrialism and dominant themes in a different area of physics prominently cultivated by British physicists, thermodynamics, is considered in the classic study by D. S. L. Cardwell, *From Watt to Clausius: The Rise of*

Thermodynamics in the Early Industrial Age (Ithaca: Cornell University Press, 1971); see also the valuable study by Crosbie Smith, *The Science of Energy: A Cultural History of Energy Physics in Victorian Britain* (Chicago: University of Chicago Press, 1998).

On Faraday, the now classic biography is L. Pearce Williams, *Michael Faraday: A Biography* (New York: Basic Books, 1965). Regarding Faraday's attitude towards the reality of field lines, Ian Hacking, *Representing and Intervening: Introductory Topics in the Philosophy of Natural Science* (Cambridge: Cambridge University Press, 1983), provides an interesting defense of belief in the reality of entities that are conceptualized as tools or instruments; on similar issues, with specific discussion of Faraday, see also David Gooding, "'In Nature's School': Faraday as an Experimentalist," in *Faraday Rediscovered: Essays on the Life and Work of Michael Faraday, 1797–1867*, ed. David Gooding and Frank A. J. L. James (London: Macmillan, 1985), pp. 106–35; Gooding, *Experiment and the Making of Meaning: Human Agency in Scientific Observation and Experiment* (Dordrecht: Kluwer, 1990). The relationship between British acceptance of Faraday's field-line ideas and their apparent practical instantiation in telegraphy is examined in Bruce J. Hunt, "Michael Faraday, Cable Telegraphy, and the Rise of Field Theory," *History of Technology* 13 (1991), pp. 1–19; Hunt, "Doing Science in a Global Empire: Cable Telegraphy and Electrical Physics in Victorian Britain," in *Victorian Science in Context*, ed. Bernard Lightman (Chicago: University of Chicago Press, 1997), pp. 312–33. On Faraday's and other theoreticians' relationships to practical "electricians," see also Iwan Rhys Morus, *Frankenstein's Children: Electricity, Exhibition, and Experiment in Early-19th-Century London* (Princeton: Princeton University Press, 1998).

Finally, a study of Maxwell's use of both analogy and mechanical models in his physics, examined as forms of metaphor, is "Maxwell et la 'métaphore scientifique,'" chap. 9 of Fernand Hallyn, *Les structures rhétoriques de la science: De Kepler à Maxwell* (Paris: Éditions du Seuil, 2004).

Chapter Six

The two most influential works for the approach taken in this chapter are Mara Beller, *Quantum Dialogue: The Making of a Revolution* (Chicago: University of Chicago Press, 1999), and James T. Cushing, *Quantum Mechanics: Historical Contingency and the Copenhagen Hegemony* (Chicago: University of Chicago Press, 1994).

Accessible works on early quantum physics that provide more than just impressionistic accounts of the intellectual content are rare; an exception is Bruce R. Wheaton, *The Tiger and the Shark: Empirical Roots of Wave-Particle Dualism* (Cambridge: Cambridge University Press, 1983). Also of value is Helge Kragh, *Quantum Generations: A History of Physics in the Twentieth Century* (Princeton: Princeton University Press, 1999), esp. chaps. 4, 5, 11, and, for the rise of the Copenhagen interpretation, chap. 14. For a survey of the earlier history, see Martin J. Klein, "The Beginnings of the Quantum Theory," in *History of Twentieth Century Physics*, ed. Charles Weiner (New York: Academic Press, 1977), pp. 1–39; of the more obviously demanding treatments of relevant issues, see Thomas S. Kuhn, *Black-Body Theory and the Quantum Discontinuity, 1894–1912,* 2nd ed. (Chicago: University of Chicago Press, 1987), which also differs somewhat from Klein's approach; Olivier Darrigol, *From c-Numbers to q-Numbers: The Classical Analogy in the History of Quantum Theory* (Berkeley: University of California Press, 1992).

Regarding terminology in this period, it is worth remembering that the terms "classical mechanics" and "classical physics" referred to little more than the mechanics or physics that was second nature to contemporary physicists from their own professional training, and in which energy was always continuous, not quantized. A useful biography of Bohr is Abraham Païs, *Niels Bohr's Times, in Physics, Philosophy, and Polity* (Oxford: Clarendon Press, 1991), which also provides quite accessible accounts of the development of quantum physics from Planck onwards. A more popular, nontechnical biography is Niels Blaedel, *Harmony and Unity: The Life of Niels Bohr* (Madison: Science Tech Publishers, 1988).

There are, of course, many studies dealing with Albert Einstein. Aspects of Einstein's methodological ideas near the beginning of his career are valuably discussed in Gerald Holton, "Mach, Einstein, and the Search for Reality," in Holton, *Thematic Origins of Scientific Thought* (Cambridge, MA: Harvard University Press, 1988), pp. 237–77. Among the better biographies are Abraham Païs, *Subtle Is the Lord: The Science and the Life of Albert Einstein* (Oxford: Clarendon Press, 1982), and a brief, nontechnical, but serious work by David C. Cassidy, *Einstein and Our World* (Atlantic Highlands, NJ: Humanities Press, 1995). Cassidy is also the author of a well-regarded biography of Werner Heisenberg: David C. Cassidy, *Uncertainty: The Life and Science of Werner Heisenberg* (New York: W. H. Freeman, 1992). The most relevant to his natural philosophy of Einstein's popular essays are collected in Albert Einstein, *Ideas and Opinions* (New York: Crown Publishers, 1954; with frequent reprints). See also a classic collection of reminiscences and appreciations by his contemporaries near the end of his life: Paul Arthur Schilpp, ed., *Albert Einstein: Philosopher-Scientist* (Evanston: Library of Living Philosophers, 1949). Arthur Fine, *The Shaky Game: Einstein, Realism, and the Quantum Theory* (Chicago: University of Chicago Press, 1986), is an important historical and philosophical study of Einstein's attitudes and arguments as they related to quantum mechanics.

Paul Forman's classic article on quantum acausality and German cultural malaise is Forman, "Weimar Culture, Causality, and Quantum Theory, 1918–1927: Adaptation by German Physicists and Mathematicians to a Hostile Intellectual Milieu," *Historical Studies in the Physical Sciences* 3 (1971), pp. 1–115, while among articles contesting or criticizing Forman's argument are John Hendry, "Weimar Culture and Quantum Causality," *History of Science* 18 (1980), pp. 155–80, and P. Kraft and P. Kroes, "Adaptation of Scientific Knowledge to an Intellectual Environment: Paul Forman's 'Weimar Culture, Causality, and Quantum Theory, 1918–1927': Analysis and Criticism," *Centaurus* 27 (1984), pp. 76–99.

On the Einstein-Podolsky-Rosen paradox (EPR), see, as well as Fine, *Shaky Game*, chap. 3, Mara Beller and Arthur Fine, "Bohr's Response to EPR," in *Niels Bohr and Contemporary Philosophy*, ed. Jan Faye and Henry J. Folse (Dordrecht: Kluwer, 1994), pp. 1–31. For a very different kind of analysis, see David Kaiser, "Bringing the Human Actors Back on Stage: The Personal Context of the Einstein-Bohr Debate," *British Journal for the History of Science* 27 (1994), pp. 129–52. (EPR's essence was fully developed only after the advent of Bell's theorem and ideas of nonlocality in the 1960s, which restructured the assumptions that were at issue.) Schrödinger's "cat paradox" is discussed in Fine, *Shaky Game*, chap. 5.

On David Bohm and his opposition to the Copenhagen interpretation: besides Cushing, *Quantum Mechanics*, chaps. 4, 9, and Beller, *Quantum Dialogues*, pp. 206–10, compare David Bohm, *Causality and Chance in Modern Physics* (New York: Harper Torchbooks, 1961 [1957]), with Werner Heisenberg, *Physics and Philosophy* (London: Penguin Books, 1989 [1962]). On Bohm and the von Neumann "proof," see also Trevor J. Pinch, "What Does a Proof Do If It Does Not Prove? A Study of the Social Conditions and Metaphysical Divisions Leading to David Bohm and John von Neumann Failing to Communicate in Quantum Physics," in *The Social Production of Scientific Knowledge*, ed. Everett Mendelsohn, Peter Weingart, and Richard Whitley (Dordrecht: Reidel, 1977, pp. 171–215. An excellent collection of nontechnical interviews with physicists, including Bohm, is P. C. W. Davies and J. R. Brown, eds., *The Ghost in the Atom: A Discussion of the Mysteries of Quantum Physics* (Cambridge: Cambridge University Press, 1986), which also contains a good introduction by the editors to EPR, Schrödinger's cat, Bell's theorem, and other aspects of contemporary understandings of quantum physics: "The Strange World of the Quantum," pp. 1–39.

Conclusion

For more on the aesthetic aspects of evaluation in science, see James W. McAllister, *Beauty and Revolution in Science* (Ithaca: Cornell University Press, 1996); also Arthur I. Miller, *Insights of Genius: Imagery and*

Creativity in Science and Art (New York: Copernicus/Springer-Verlag, 1996). Boris Castel and Sergio Sismondo, *The Art of Science* (Peterborough, Ontario: Broadview, 2002), takes up in addition the idea of sciences as practical crafts, to some extent reflecting the "instrumentality" aspect of science discussed in the present book; see also, on related themes, Pamela H. Smith, *The Body of the Artisan: Art and Experience in the Scientific Revolution* (Chicago: University of Chicago Press, 2004).

Boyle's notion of the "priest of nature" is classically emphasized in Harold Fisch, "The Scientist as Priest: A Note on Robert Boyle's Natural Theology," *Isis* 44 (1953), pp. 252–65.

Kant and Kantianism are studied in Michael Friedman, *Kant and the Exact Sciences* (Cambridge, MA: Harvard University Press, 1992). Positivism in chemistry in the late nineteenth and early twentieth centuries is examined in Mary Jo Nye, *Molecular Reality: A Perspective on the Scientific Work of Jean Perrin* (London: Macdonald, 1972), and Nye, *Before Big Science*, chap. 4, with a more general overview of positivistic attitudes among physical scientists in the classic essay by John L. Heilbron, "Fin-de-siècle Physics," in *Science, Technology, and Society in the Time of Alfred Nobel*, ed. Carl-Gustav Bernhard, Elisabeth Crawford, and Per Sèrböm (Oxford: Pergamon Press, 1982), pp. 51–71; see also Rocke, *Chemical Atomism* (listed above for chapter 3). On Poincaré and "conventionalism," see Peter Louis Galison, *Einstein's Clocks, Poincaré's Maps: Empires of Time* (New York: Norton, 2003). On philosophy of science in the earlier twentieth century, see Alan Richardson, "Toward a History of Scientific Philosophy," *Perspectives on Science* 5 (1997), pp. 418–51; among many introductions to twentieth-century approaches to the philosophy of science, see A. F. Chalmers, *What Is This Thing Called Science?*, 3rd ed. (Indianapolis: Hackett, 1999).

Michael Aaron Dennis, "Reconstructing Sociotechnical Order: Vannevar Bush and US Science Policy," in *States of Knowledge: The Co-Production of Science and Social Order*, ed. Sheila Jasanoff (London: Routledge, 2004), pp. 225–53, provides invaluable background to the development of the U.S. scientific establishment and its values following the Second World War.

For the context of the earl of Newcastle's group around 1600, see Robert Kargon, *Atomism in England from Hariot to Newton* (Oxford: Clarendon Press, 1966); John Shirley, *Thomas Harriot: A Biography* (Oxford: Clarendon Press, 1983); Robert Fox, ed., *Thomas Harriot: An Elizabethan Man of Science* (Aldershot, UK: Ashgate, 2000). Broader discussion of the issues relating to alchemy and practical knowledge in premodern Europe may be found in William R. Newman, *Promethean Ambitions: Alchemy and the Quest to Perfect Nature* (Chicago: University of Chicago Press, 2004).

On the anthropic cosmological principle: John D. Barrow and Frank J. Tipler, *The Anthropic Cosmological Principle* (Oxford: Clarendon Press, 1986), is the major statement, with historical discussion in chaps. 2 and 3; see also the earlier, brief account in Paul Davies, *Other Worlds: A Portrait of Nature in Rebellion. Space, Superspace and the Quantum Universe* (New York: Simon and Schuster, 1980), chap. 8. A useful essay collection is F. Bertola, and U. Curi, eds., *The Anthropic Principle: The Conditions for the Existence of Mankind in the Universe* (Cambridge: Cambridge University Press, 1993). The principle is often discussed nowadays in connection with theistic arguments concerning "intelligent design," where its main use is as a source of apparently designful features of the world: see, e.g., Michael A. Corey, *The God Hypothesis: Discovering Design in Our "Just Right" Goldilocks Universe* (Lanham, MD: Rowman & Littlefield, 2001).

On Einstein and "theories of principle," see Holton, "Mach, Einstein, and the Search for Reality."

For an introduction to the ancient skeptical philosophy of Sextus Empiricus, see Richard H. Popkin, *The History of Scepticism from Savonarola to Bayle* (Oxford: Oxford University Press, 2003), chap. 2. On mathematical proofs and computers, see the popular treatment in Simon Singh, *Fermat's Last Theorem* (London: Fourth Estate, 1997), and especially the studies by Donald MacKenzie, *Knowing Machines: Essays on Technical Change* (Cambridge, MA: MIT Press, 1996), esp.

chap. 8, and MacKenzie, *Mechanizing Proof: Computing, Risk, and Trust* (Cambridge, MA: MIT Press, 2001). Larry Wos and Gail W. Pieper, *Automated Reasoning and the Discovery of Missing and Elegant Proofs* (Paramus, NJ: Rinton Press, 2003), makes the case for the value of computer-assisted proofs and provides examples.

Various versions of "theory/experiment" views of science include Stephen Toulmin, *Foresight and Understanding: An Enquiry into the Aims of Science* (New York: Harper Torchbooks, 1963 [1961]); Ian Hacking, *Representing and Intervening: Introductory Topics in the Philosophy of Natural Science* (Cambridge: Cambridge University Press, 1983); also a more complex examination of similar issues by Peter Galison, *Image and Logic: A Material Culture of Microphysics* (Chicago: University of Chicago Press, 1997).

On computers and chess, a useful, if now rather dated, discussion of the principles of computer chess for the nonspecialist is David Levy and Monty Newborn, *How Computers Play Chess* (New York: Computer Science Press, 1991). For "God endgames," see Tim Krabbé's "Open Chess Diary," at http://www.xs4all.nl/~timkr/chess2/diary.htm, diary entry #60, with additional references.

Popular views in the 1920s and '30s of Einstein's work are examined in Abraham Païs, *Einstein Lived Here* (Oxford: Clarendon Press, 1994), esp. chap. 11, while the basic introduction to string theory for the nonphysicist is Brian Greene, *The Elegant Universe: Superstrings, Hidden Dimensions, and the Quest for the Ultimate Theory* (New York: Norton, 1999).

The literature on non-Western natural philosophies and technologies, as well as on science and colonialism, is large and expanding rapidly. The most intensive large-scale study of the former is Joseph Needham et al., *Science and Civilisation in China*, 7 vols. (Cambridge: Cambridge University Press, 1954–2004), while Geoffrey Lloyd and Nathan Sivin, *The Way and the Word: Science and Medicine in Early China and Greece* (New Haven: Yale University Press, 2002), is a comparative study of ancient approaches in two different cultural settings by leading historians of ancient Greek and ancient/medieval Chinese natural philosophy respectively. A useful starting point on relevant colonial issues is Roy MacLeod, ed., *Nature and Empire: Science and the Colonial Enterprise*, a volume of the annual periodical *Osiris*, n.s. vol. 15 (2000), while a challenging recent study is Gyan Prakash, *Another Reason: Science and the Imagination of Modern India* (Princeton: Princeton University Press, 1999).

Index